Fort Nelson Public Library
Box 390
Fort Nelson, BC
V0C-1R0

Copyright © 2002 by Westcoast Energy Inc.

02 03 04 05 06   5 4 3 2 1

All rights reserved. No part of this book may be reproduced, stored in a retrieval system or transmitted in any form or by any means without the prior permission of the publisher or, in the case of photocopying or other reprographic copying, a licence from CANCOPY (Canadian Copyright Licensing Agency), Toronto, Ontario.

Douglas & McIntyre
2323 Quebec Street, Suite 201
Vancouver, British Columbia V5T 4S7

National Library of Canada Cataloguing in Publication Data
Newman, Peter C., 1929–
Continental reach
Includes index
ISBN 1-55054-969-3

1. Westcoast Energy Inc.—History. 2. Gas industry—Canada—History. 3. Gas companies—Canada—History. I. Title.
HD9581.C34W47 2002   338.7′6223385′0971   C2002-910279-0

Editing by Brian Scrivener
Research and editing by Frank Dabbs
Jacket and text design by Peter Cocking
Jacket photograph by Steve Short
Typesetting by Rhonda Ganz
Printed and bound in Canada by Friesens
Printed on acid-free paper

PETER C. NEWMAN

REACH

the WESTCOAST ENERGY *story*

Douglas & McIntyre
Vancouver/Toronto

Continental Reach

Fort Nelson Public Library
Box 330
Fort Nelson, BC
V0C-1R0

CONTINENTAL

Contents

Introduction *1*
Prologue *3*

1. The Founding Father *7*
2. The Wildcat Dreamer *17*
3. The Peace Country Project *29*
4. The Pipeliners *41*
5. Building the Line *52*
6. The Force of Will *63*
7. The Leadership Crisis *71*
8. The Great Conciliator *84*
9. The Conciliator's Brand *96*
10. The Transition *109*
11. The Steward *125*
12. The Trailblazer and His Team *134*
13. The Eastern Breakout *151*
14. The Agenda *167*
15. Transforming Corporate Culture *179*
16. The North American Footprint *190*
17. Mexico *203*
18. The Canadian North *212*
19. The Cornucopia *225*
20. On the Threshold *232*
21. The Deal of a Lifetime *243*

Index *253*

Introduction

This is the history of a company that began as a map scribbled on a scrap of paper by the inferno light of a runaway well, charting a wilderness pipeline from natural gas fields that existed in the mind of a frontier wildcatter to an undeveloped market where people split their own firewood and stoked their furnaces with coal. It was an unlikely notion, and the dreamer who drafted the sketch had spent his life chasing rainbows that never led to a pot of gold. The idea was ambitious because no one had built a pipeline quite like it over the untamed terrain of Canada's sub-arctic solitude.

To the surprise of everyone except the founder, the idea succeeded. Endowed with more engineering zeal than public relations imagination, the resulting company called itself Westcoast Transmission, which prompted most puzzled observers to think of it as an automobile repair business, fixing bum clutches.

This is the story of a sometime sleepy company that burst the boundaries of its self-circumscribed Pacific Coast world and renamed itself Westcoast Energy. It went exploring with the vigour of the legendary coureurs de bois until the continental reach of its operations and interests ran from Canada's Pacific shoreline to its Atlantic basins and Mexico's Campeche Bay to Alaska's Prudhoe Bay. It was as if Woody Guthrie had written his immortal people's anthem, "This Land Is Your Land," with this company in mind.

This is the story of a company that found itself, at the end of its century, a $15 billion corporation competing in a $30 billion world. Its choice was to stagnate or to reward the faith of generations of investors by finding a home big enough to absorb its North American footprint. Westcoast found that place in Duke Energy, a company from the Carolina states on the other side of the continent, one with a history and culture uncannily like its own.

Too busy realizing its future, Westcoast was never mesmerized by its past. It was always more interested in making profit than in making history. Not that its story has lacked fascination. From its origins, when deals were struck with a handshake after shots of whiskey, Westcoast diversified in many directions and several continents, reaching the top rank of Canadian corporations in terms of operational and financial size and becoming one of only a handful in terms of a story that became a Canadian corporate legend.

The Westcoast tale is as absorbing as any on the too-thin shelf of Canada's corporate histories. It deserves to be told because even if gas and oil pipelines are an essential part of the fabric and marrow of Canadian life, they do not appear on most citizens' radar screens. They cast no shadow. Their chief asset is buried below ground, mostly in that part of Canada's mysterious hinterland the poet Al Purdy defined as being "north of summer."

Westcoast's spirited, daring progress makes its evolution a compelling tale. Embedded in a great North American company named Duke Energy, there will always be the legacy of Canadian enterprise, born in the northern vastness, shaped on the Western landscape, tested in every corner of the continent and engaged in making the twenty-first century North America's.

Prologue

Near the village of Pouce Coupe in 1921, a well struck gas, blew out and burned for many years. This well produced from sandstone that appeared to be the same as in the vast Viking field near Edmonton. This parallel gave me the idea that there might be another extensive field like the Viking in the Peace River area, which could serve as a supply for Vancouver, some 650 miles to the south.

FRANK McMAHON

It was the bear pit of the Great Depression and no one—except his apprehensive investors back in Vancouver—was paying attention to the travels and travails of a short, stocky wildcatter named Frank McMahon. His obsessive hunt for oil-bearing formations along the eastern apron of the Rocky Mountains had taken him to this forlorn campsite deep in the forested hills of the fabled Peace River country, close to the Alberta–British Columbia border. As he cooked a modest supper, the road-weary McMahon, who had lost a couple of fortunes but never his sense of humour, laughed to himself and likened his lot to that of the hapless Quebec trapper who had cut off his thumb with an axe, giving this otherwise undistinguished district its name.

Then he saw it.

It was dusk, yet there was light in the eastern sky.

Near the bank of the Pouce Coupe River, on the shore opposite his encampment, a tall thin spike of flame cast a pantomime of light and shadow on canyons of pine and poplar. As the blaze danced across the water, McMahon realized that the steady roar he could hear was not that of the river gone wild but of an untamed inferno.

Although he was familiar with the flares of oil wells from Turner Valley to Texas, McMahon had never witnessed nature's fireworks quite so spectacularly. He knew the source of this earthbound aurora. In this trackless land of Ice Age aboriginal campsites, eighteenth-century fur traders and nineteenth-century homesteaders, McMahon had found the source of a twentieth-century legend he had heard about in the oil patch watering holes of Montana and Alberta. The source of the spectacle was a wildcat well drilled by the swashbuckling geologists of Imperial Oil during their original search for the Canadian petroleum basin's motherlode. They had abandoned their digging near the outpost of Pouce Coupe but not before tapping into a sandstone natural gas reservoir that had accidentally run wild and caught fire.

At the time, Calgary had been heating its homes with natural gas for two decades, and Rudyard Kipling had visited the town of Medicine Hat, where he coined the phrase "all hell for a basement" to describe its gas-fired economy of pottery kilns, street lights and house furnaces. Alberta's micro-markets for natural gas were an accident of geography. Oil promoters and wildcatters had little interest in natural gas. They regarded the stuff as a nuisance by-product for which there was no current market of worthwhile scale and little prospect of any in the foreseeable future. No one knew how to build natural gas pipelines big enough or long enough to link remote gas fields with major cities. Gas was held in such contempt that burning it off at Turner Valley to produce oil created enough light by which to read a newspaper at night and bums slept by the flares on cold winter nights in the Depression.

Yet the homesteaders of rural British Columbia still split wood for their kitchen stoves, and the citizens of the Lower Mainland shovelled cinders from their coal furnaces or paid a small fortune to fuel their oil burners. Frank McMahon, who had helped his mother bring in wood and had sold those oil burners house to house in Vancouver, sat there on the bank of the river watching the mesmerizing display, speculating that he'd stumbled onto something close to an eternal flame. How many men had sat on how many nights watching this beacon without realizing that under their feet lay a huge, untapped reservoir of energy? Here, McMahon believed, was a prime fuel to heat the homes and buildings of Vancouver, Seattle and San Francisco, and power for the industries of those cities—the flame that could percolate their coffee, cook the gourmet meals of their finest restaurants and fuel their vehicles.

That night, McMahon visualized two maps. He scratched them out on a scrap of paper to show his partners and investors. The first map was the geological profile that connected the gas-bearing sandstone of Pouce Coupe with the reservoir rock of the biggest natural gas fields in Alberta. McMahon, who had debated his ideas with some of the best geologists of the day, believed these reservoirs belonged to the same geological formation that extended into northeast British Columbia's Peace River country. The second chart showed a thin pencil line from this sandstone storehouse across the rugged heart of British Columbia to Vancouver—and south into the United States. This was the right-of-way for a natural gas pipeline the like of which had never been built, and which, as far as McMahon was aware, no one knew how to construct.

It took the better part of a quarter-century to realize the dream of that pipeline, initially glimpsed in the fire of the wild well on the banks of the Pouce Coupe on that summer's night in 1934. Bringing it to life turned the obscure, penniless dreamer into a rich and worldly individual who made history and moved markets. And yet as remarkable as that first pipeline achievement was, it was just the kindling spark that gave life to Westcoast Energy Inc.

I

The Founding Father

They could not have come from anywhere but Canada.
Something in their bearing told the story—a naturalness, a breeziness,
an alertness which suggested the new world. Self-control, an air
of discipline and good manners. They were rarely found
lounging; they seemed always to have some purpose in mind.

THE HON. VINCENT MASSEY

Frank McMahon was a man of his times and his times never stopped changing. During his formative years, his country's future was being defined not in the banks and industrial plants of the cities but on rural homesteads and land grants, in the timber forests and waterways of its inscrutable wilderness, and at the muddy drilling rigs and mining camps where men moiled for gold and other treasures of the earth.

There is a storybook quality to his life as one of three sons produced during the four-year marriage of a teenage schoolteacher and a boomtown drifter and bootlegger seeking his fortune in Western Canada's mining camps. Francis Joseph McMahon and his cardsharp brother Patrick started their prospecting and poker careers in the Klondike gold rush of 1898. Two years later, they had drifted south to do some silver and lead prospecting at Moyie, in British Columbia's wild East Kootenay district. From the Wildhorse Creek gold field at Fort Steele to the St. Eugene silver lode on Yahk Mountain, a few kilometres south of Moyie, this was the Rocky Mountains' hottest mineral play, rivalling the rush to riches in Ontario's Haliburton highlands. The McMahon brothers purchased one of Moyie's six hotels, providing the income that allowed Francis to scour the lake-dotted mountains and valleys for unstaked claims, while Patrick stayed home and played high-stakes poker. Most of the breadwinners in Moyie, a town of five hundred shivering souls, worked the St. Eugene mine or in the nearby logging camps. It was a hardscrabble world, a man's domain of mud-spattered labour relieved by drinking, playing keno and whoring in the hotel saloons that never closed.

Bush settlements like Moyie are remembered for their frontier gypsies like Francis McMahon, but there were other, more stable, citizens, such as migrant clergymen and threadbare remittance men surviving on small emoluments, who built log cabins and clapboard chapels. The professional families sponsored amateur theatrical societies and other cultural amenities, while the affluent townsfolk of the nearby railway depot of Cranbrook developed a resort on Moyie Lake. Its comfortably appointed cottages and lodge boasted a flotilla of small sailboats and the bucolic pleasures of casting for the six species of trout that amply inhabited local waters.

One of the little community's civilizing elements was Stella Soper, the daughter of a retired army major turned railway bridge contractor who had helped the Canadian Pacific Railway (CPR) construct a web of branch lines hauling East Kootenay ore production to the railway's smelter at Trail. Stella's parents owned the North Star Hotel in nearby Kimberley, but she often came south to Moyie to teach music in the school and give private lessons to children whose parents wanted refinement to counter the town's harsh frontier atmosphere. Following a whirlwind courtship, Francis McMahon married Stella in 1901; their first son, Francis (Frank) Murray Patrick, was born October 2, 1902. Two more boys followed: George in 1904 and John in 1905. At about this time, the St. Eugene strike played itself out and no new discoveries had presented themselves. Francis had heard rumours of gold found north, at Barkerville in the Columbia Mountains. The McMahon brothers wanted to move on; Stella and the boys did not. The couple parted. If there was acrimony, she kept it from the children, but the plain fact was that Francis walked out on his family and rarely contacted them again.

Had the McMahon family stayed together, Frank and his brother George, who became business partners, would have been just another couple of rootless prospectors and wildcatters. Western Canada was replete with such men in the first half of the twentieth century, rugged outcasts who never developed the business acumen to go beyond accumulating, with luck, minor fortunes and major hardships. Frank McMahon and his brothers were not wanted on the journey with their father and uncle. So their upbringing passed into the hands of their mother, their grandmother Mary and, later, the family friend who became Stella's second husband, Vancouver businessman and mining engineer Owen C. Thompson. Under this aegis, the boys received the education, the social breeding and the intangible senses of possibility and direction that equipped them to turn their ambitions into the lasting achievements denied their father and uncle.

Frank's impressionable preschool years in Moyie at the height of its mining boom and his father's restless Irish blood whetted his taste for the risks that he took exploring for oil. By his own account, these provided him the motive for building a successful natural gas production and transportation empire. He wanted, he later recalled, to give people the kind of comfort and convenience of daily living denied his mother in the rough-and-ready cabin where they survived during his first five years. "Your mother and father were pioneers and my mother and father were pioneers," he once told his close friend and aide Pat Bowsher. "They always had a woodpile outdoors. I want to make it possible for mothers not to have to go out to the woodpile, but just to turn on the gas, right across British Columbia."

Frank McMahon's education was not limited to the romance of the manure-carpeted streets of a mining camp. It was the product of a more enduring sense of discipline and perseverance garnered from his mother's determination to raise her boys to a higher level—the quintessential twentieth-century dream of a mother for her children to have a life better than hers.

After her separation, Stella remained in Moyie, but the town was dying. Her mother, now widowed, was living comfortably on the income from the North Star Hotel in Kimberley. When Frank was five, the family moved there to help manage the hotel. The mineral wealth that fuelled East Kootenay's growth had largely eluded Kimberley until the First World War. When Stella and her boys arrived, the hamlet was little more than a supply stopover for prospectors and miners. The North Star Hotel, named after the local North Star mine, was its only inn. When the boys started school, classes were held in the hotel. The McMahons were happy there, growing up outdoors in a world of sand-pit baseball games, buddy dogs, campfire roasts, rockhounding on the mountainsides, swimming in summer and skating in winter. In the evenings, Stella taught them to love music, appreciate good books and take a lively interest in community life, from politics and business to live theatre and Saturday night community dances. One of the North Star Hotel's residents was Owen C. Thompson, a mining engineer who had come to Kimberley to invest in its resources but wisely maintained himself on his professional income as consultant and business manager while waiting for his capital to pay its dividends. Thompson took an interest in the McMahon boys while courting their mother.

Frank owed the completion of his education not only to his mother's resolve that he succeed, but also to the good offices of the local Presbyterian church. There was no adequate high school in the East Kootenays. After he finished primary school, Stella sent him to Calgary, where the newly opened Western Canada High School took rural students. It was developing a reputation as an egalitarian finishing school for families in southwestern Alberta and southeastern B.C. who wanted more education for their children than local schools could provide. During Frank's first year away from his mother's firm hand, the lanky, tousle-headed kid became a superb collegiate athlete and a freewheeling spirit. Popular and gregarious, he attached himself to a circle of affluent youngsters more interested in dancing, boozing and fast automobiles than studies. After a year of more hijinks than homework, it became evident that he needed a firmer hand to complete his schooling.

Stella conferred with Owen Thompson and the clergy. She looked south to the private parochial schools of the Pacific Northwest. The road to Frank McMahon's ambition ran south naturally enough, because the towns of the Kootenay region faced in that direction. Vancouver was a long 520 miles to the West through then-impassable mountains. Spokane was just 190 miles away, no more than a day's drive. In the autumn of 1917, Frank McMahon was admitted to Whitworth College, a private Presbyterian preparatory school in Spokane, Washington. The college founder was a religious utopian named George Whitworth who had migrated to the Oregon Territory in the mid-1800s to found a Presbyterian academy that would provide "a good English education and a thorough religious training" for promising young men and women west of the Mississippi. When Frank McMahon arrived for classes, the faculty still aspired to produce "morally guarded" scholars, but had opened its doors to non-Presbyterians. "While Whitworth College is denominational, it does not aim to be sectarian, opening its doors to all lovers of truth and learning," its brochures boasted.

If his wild term at Western Canada High School prepared him for the rambunctious reputation he was to earn in business, Whitworth provided Frank McMahon with his best years as a serious student. The liberal arts program, which later elevated the college to its sterling status as a reputable post-secondary degree-granting institution, struck a chord in the young man's mind and spirit. His teachers stimulated his expanding intellect. Their discipline kept him focussed on his studies. He received his moral and ethical grounding here and learned to esteem the worlds of ideas and science. He turned his boisterous, unruly humour into a sense of wit and grace that served him when his business career eventually took him to the highest levels of political, financial and social life in New York, Hollywood, London and Washington. He got to know himself well enough to recognize that he would never have the patience for great academic accomplishments, but he gained respect for the discipline and knowledge of the professionals with whom he later worked as a business executive.

He was also taught to play baseball at a school where athletics was considered serious business and professional scouts came calling each year to look over the prospects the coaches had recommended. Quick for his size, and capable of a fierce level of concentration, McMahon became a varsity-calibre shortstop and catcher, as well as a pretty good batter. He loved the smell of lanolin oil on leather gloves and the dust of the infield. He enjoyed being part of the team, but most of all he loved to win. Back home in Kimberley each summer, he laboured on road crews, in the mines, in the sawmills. In the evenings and on weekends, he played ball for any team that would have him, as far away as Cranbrook.

His mother discouraged the talk around town and at the dinner table of his playing baseball for a living. She could visualize him, easygoing as he was, drifting into the never-never land of the semi-pro leagues where he wouldn't notice his life slipping by. Things came so easily to him she feared that he wouldn't have the drive to push for the top and that he'd have a good enough time in the boozy, one-night life of the Triple A's and might even grow content. It was a world too similar to the one that had sucked his father and uncle into mediocrity. She heard whispers of news about Francis. That he drifted in and out of the hotel business, moving from one British Columbia mining town to another. That he went broke in pre-war land development in the Peace. That he was in Coalmont, presiding over the local hostelry where the miners talked trade union politics late into the night and drank themselves under the tables. Stella did not want her son to drift in similar fashion, from baseball town to baseball town along the red-clay roads of Alabama or the sage-lined byways of Texas.

When Frank McMahon graduated from Whitmore, he signed up for business administration classes at Gonzaga University, a private Jesuit college across town. The priests at Gonzaga followed the 450-year-old Ignatian ideal of educating the whole student, "mind, body and spirit in an integration of science and art, faith and reason, action and contemplation." But they got their hands on Frank McMahon too late. For two more winters he cracked the books, and for two more summers, he sweated at his pick-up jobs in Kimberley and played more baseball. But he was unhappy at Gonzaga and couldn't see much point in being there. It wasn't the management of business that interested him; he had no ambition to become a corporate bureaucrat. He wanted to create enterprises, to make money. Someone else could do the paperwork.

McMahon spent less time with his books and more time on the social circuit. As usual, he gravitated to the active and most-interesting students, including a popular Spokane youngster named Harry Lillis Crosby, who was turning a profit on his college education by singing at campus dances. "Bing" Crosby was a practising Catholic who wholeheartedly embraced the Jesuitical philosophy, but he also relished the raucous social life that the onset of the Roaring Twenties brought to Gonzaga—and there he found lifelong common ground with McMahon's Irish sociability.

DURING HIS THIRD year at university, Frank McMahon quit to make a living. He had a singular ambition. He declared: "I'm going to be a millionaire." Further education seemed an impediment, but his first job—as an unsuccessful door-to-door salesman on the streets of Spokane, peddling gasoline coupons for Standard Oil (later to become Chevron)—was a disaster. He made little money and, when summer returned, parlayed his summer labouring experience at the Canadian Pacific Railway–owned Consolidated Mining and Smelting Co. (later Cominco) Sullivan lead zinc mine near Kimberley into a permanent job on a crew searching for the Sandon silver deposits at Slocan in the East Kootenays. He was promoted quickly to be the diamond setter and boss driller of his crew. Mitchell Diamond Drilling Co. was a contract driller and for the next three years McMahon and his men worked up and down the Pacific Northwest, from San Francisco to Alaska, on mining and construction jobs, even drilling footings for San Francisco's

Bay Bridge. (Out of that trip came the family legend that he had drilled footings for the Golden Gate Bridge, but construction on that engineering behemoth did not commence until 1933, and by then Frank McMahon had made a career change and was drilling for oil and gas in Montana and Alberta.)

Owen Thompson, meanwhile, had prospered. When its operator abandoned the North Star mine in 1918, he leased the property and for several seasons successfully salvaged profitable quantities of ore from its tailings and surface digs. Stella obtained a divorce and she and Owen married, then moved to Vancouver, where Frank's brothers, George and John, worked as investment salesmen peddling bonds. Thompson soon became a leading figure in the financing and planning of mining ventures.

When Frank was twenty-four years old, Mitchell Diamond Drilling sent him to open a Vancouver office and find business for the company. He soon grew restless with his desk-bound job and took off for California, where he worked again for Standard Oil, this time selling bulk contracts to industrial consumers. But his stepfather urged him back to Vancouver, where they spent many hours together as the older man showed him how to take his savings and invest them in his own independent drilling company with a down payment on three rigs and bank backing for the balance. Frank negotiated his first diamond-coring mining exploration contract and parted from Mitchell on such good terms that they sold him his first rig. From its start-up, his venture did well, as his Pacific Coast mining connections and his good reputation paid off.

His largest assignment was to drill the dynamite holes for a twenty-five-mile tunnel from the Hetch Hetchly water reservoir through to the Coast Mountains to San Francisco. In 1928, following the aqueduct's completion, there was an ominous slowdown in construction and mining work in the West. The Roar was dissipating from the Twenties, though McMahon first took it merely as one of the periodic cycles in the exploration business. He returned to Vancouver and took a temporary position selling oil furnaces door-to-door. He hated the job; he was tongue-tied and ill at ease with the housewives who answered his knock. But he learned the lesson that changed his life: coal, the fuel of choice for the domestic furnaces of Vancouver, was on its way out— along with the dust and clinkers that every homeowner hauled out of the fire bed each weekend to scatter in the driveway. Oil, shipped by tanker from the refineries of California, was comparatively clean and convenient. In Alberta, at a place called Turner Valley, wildcatters were getting rich in a second oil boom (the first having taken place just before the First World War). There were accounts of oil seeps in the Fraser River delta, and it seemed just a matter of time before Canadians would strike their own black gold.

McMahon might have embarked on his wildcatting career then and there, but his mind was otherwise occupied. George had just introduced him to his girlfriend, Jessie Grant. And Jessie had a sister—a peaches-and-cream blond beauty named Isobel. She was a university dropout and secretary who held down a second job working evenings in the records department of the Vancouver Stock Exchange and under George's tutelage she was making a tidy sum investing in commodities. In a Force 10 courtship that recalled Francis McMahon's pursuit of Stella Soper twenty-eight years previously, Frank spent the summer at beach parties and bridge games, seeking Isobel's consent to marry him. The two were a couple from the photogravures of the 1920s. His smart suits and fashionable California bowler did little to disguise his well-proportioned, muscular physique. He was well off, owned a successful business and knew his way around the mining and financial world of the city. His mother and stepfather were socially well placed. Isobel was statuesque, well dressed, saucy and independent. She came from good people, loved fast cars and good parties, and she was as ambitious as he. Their wedding took place just a few rushed weeks after the couple met.

Frank didn't bother to invite his father to the occasion, and didn't raise an eyebrow when the social columns reporting the wedding claimed that Francis was deceased. The old man was, in fact, living in the mining town of Barkerville—the last old-style mining camp in British Columbia. There he ran a cigar store that depended for its survival on peddling bootleg liquor and local home-brew under the counter.

Although Frank seldom gave voice to his bitterness, he had a generally poor opinion of his father, but in spite of his hard-rock image as the rugged field hand, he was a spunky, good-natured young man. He took some comfort in knowing that his errant father had ended his reckless days in shabby dignity as the proprietor of a business that kept him fed and surrounded by the kind of men with whom he'd spent his life.

Frank and Isobel couldn't have picked a worse time to marry, considering their intention to be millionaires. For a time Frank was as busy as ever, drilling for a mining conglomerate on Vancouver Island. But after a difficult pregnancy, Isobel gave birth to their first son, Frank Jr., on the eve of the Great Depression, which abruptly ended mineral exploration, wiping out the couple's investments in stocks and commodities and bringing them to the edge of poverty. Their marriage continued for the better part of three decades, but the bitter experience of the first twenty hard years, when Isobel, scrimping for every penny, was surrounded first by diapers and then by demanding young children while Frank wandered Western Canada in search of their fortune, permanently scarred the relationship.

2

The Wildcat Dreamer

*We shall not cease from exploration
And the end of all our exploring will be to arrive
where we started
And know the place for the first time.*
T.S. ELIOT

When the 1929 crash came, Frank kept his diamond drilling company alive, at least on paper, chasing the phantoms of new contracts, and went back to selling oil burners. Neither activity could support his family, but he had one asset he could rely upon: his father-in-law's and brothers' contacts in the Vancouver financial community, including men astute enough to be surviving the worst of the Depression. McMahon believed he could persuade some of these wealthy individuals to back a venture to drill for natural gas south of the city on the geological perimeter of the Fraser delta, near the town of White Rock on the United States border. He'd gone to check out stories of gas seeps on the beach of Boundary Bay and found some aromatic, gently bubbling puddles in the tidal flats of aptly named Mud Bay. McMahon had obtained petroleum leases on several prospective drilling sites, and he proposed to drill them using his idle diamond drill rigs.

In the early 1930s, most petroleum exploration was conducted on the basis of interpreting surface contours and rock outcrops that gave clues to hydrocarbon traps deep in the ground. On the thick sediments of great river deltas, oil and gas seeps at the surface had often migrated from larger pools below. Although less precise than the infant science of geophysics and its seismological measurements, surface geology was more reliable than the folkloric methods, from divining rods to soil-tasting, used across the continent since the first, photo-finish commercial oil discoveries in Ontario and Pennsylvania in 1859. (McMahon had actually met a grizzled oil patch veteran in California in the late 1920s who claimed to know where to drill for oil because walking over the site gave him an erection.)

Frank McMahon's first wildcat partner was not one of his family contacts in the Vancouver business community, but an American promoter and pipeline operator, C. S. Shippley. McMahon had done construction-related drilling for Shippley on a primitive Missouri-to-Kansas gas pipeline that had at the time been the continent's most ambitious pipeline project. Shippley's modus operandi was to discover and claim the gas production that supplied his pipelines. He planned with McMahon to drill gas fields in upper Washington state and southern British Columbia, then build the transportation network to link the fields to Vancouver and Seattle and Portland, Oregon. Legend has it that McMahon punched $1 million worth of shallow, dry holes with the diamond drill rigs that he'd converted to rotary table oil drillers. Whether McMahon's gerrymandered rigs and tight-fisted management went through all that money, or whether most of it went into Shippley's grandiose planning and promotional fees and expenses, the outcome was the same. The sandstone reservoirs of the region may have once contained gas, but most of it had escaped before the Ice Age. In a couple of wells they did strike formations that flowed some gas, but little enough that they turned them over as a domestic fuel source to the farmers who owned the surface rights.

Shippley went back to the Midwest and McMahon returned to his Vancouver connections. Led by department-store millionaire Victor

Spencer, with Owen Thompson contributing more-modest amounts, a widening circle of Vancouverites including George Martin, Doug Boucher, Norman Guilliame, J.C. Ross, Bob Wilkinson, Norm Whittall, Frank Ross, and George and John McMahon began to back McMahon's ventures. They would eventually do extremely well, but their first gamble turned out to be an inauspicious start.

McMahon had gone back to his birthplace of Moyie, to chase down bar-room tales he had heard from his father and uncle of coal prospectors digging into pits of oil and tar. At the beginning of the Depression, he had formed a company called Crowsnest Pass Oils and, using his own funds, equipment and crew, drilled into the source rock for some oil pits dug by coal prospectors in the pass near the Alberta–B.C. border. George Hume, the Geological Survey of Canada's brilliant explorationist, had visited McMahon's drill site and encouraged him to pursue the prospect. The young promoter and the scholarly geologist shared a love for fieldwork and a passion for turning the inductive ideas of geology into economic reality. A lifelong friendship started that season. Hume became McMahon's geological mentor and, a quarter of a century later, worked for his protégé as an executive with Pacific Petroleums. Many of their discussions in the Crowsnest Pass revolved around the natural-gas-bearing sandstone formations of east-central Alberta, where drillers were bringing in fields that supplied gas to towns from Medicine Hat to Calgary. Hume was dead certain that these sandstone reservoirs were replicated along the eastern slopes of the Rockies; they would be thicker, deeper and more prolific, he speculated.

The cookhouse conversations must have been stimulating, but as the drill bits bored deeper without success, Frank ran out of funds. His reconnaissance now took him over the border to the Flathead Valley of Montana, where he acquired a handful of freehold leases from cooperative farmers. He incorporated a company called Columbia Oils Ltd. and his Vancouver sponsors staked him with $300,000 in share capital. That may have been a great deal of money in the depths of the Depression, but Columbia was as hand-to-mouth an operation as ever graced any oil patch.

Meanwhile, George McMahon was laid off from his job in the bonds business and came to work for his brother because he could type and the company needed a secretary. Frank went into the venture owing no money but having none either; the Fraser delta project had drained his bank account. He offered his drilling crew payment of three square meals a day and shares in Columbia Oils; he took the same pay package himself and went to work daily on the drilling rig floor. A photograph of that first summer on the Flathead survives. The drillers are having a picnic lunch on the rig floor with a couple of young women. Frank is all smiles and charm, a young man living a good, if impoverished, life. The gifted athlete had been reincarnated as the captain of a drilling team. The engaging socialite of college days had been sculpted by time and circumstance into an energetic, imaginative businessman. He had a frontiersman's capacity for work: in the field, he was never the biggest but often the physically strongest member of the crew. Behind his desk, he worked without much sleep or food. He was an inspired and inspiring leader but a terrible manager. George was his anchor. The brothers defined their roles early: Frank had the creative drive, flying off to new ventures, always quick to the frontier, and George sweated the small stuff.

The first well in the Flathead was a duster. Frank's backers were increasingly discouraged by their losses, and he found the accountability sessions excruciating. He was never easy asking other men to risk money and developed the habit of walking the streets, sometimes for an hour or more, to gut up for the meetings where he made his pitches. Now Frank and George McMahon were fully committed to exploring for oil and gas. Their reputations depended on it, and the Depression offered no alternative lines of work. At one point, George banged out letters to Columbia shareholders begging for $10 donations to continue drilling. Surprisingly, $200,000 came in; some shareholders contributed only $1, but most sent in $100 or more. The crew responded by promptly drilling two more dry holes.

Frank wanted to move his drilling operations to the Peace River country in northeast British Columbia. Again he was chasing oil field legends: this time, the tale of that 1921 natural gas well drilled by Imperial Oil at Pouce Coupe as part of a wide-ranging series of tests across Western Canada that eventually produced the Norman Wells oil discovery. The well had run wild and been abandoned, but Imperial's field crew left a flare stack on the wellhead to vent and burn the gas, otherwise it would have been a constant forest-fire risk. Imperial had no plans to produce the well because it was too remote for an economic pipeline. Natural gas was plentiful in Alberta and too cheap to pay for such a venture; Imperial decided there was no market with exponential growth potential to justify bringing in the well at a loss.

McMahon thought differently. He believed the Pouce Coupe wild well was producing gas from geological formations identical to those of the Viking gas field near Edmonton—the largest of natural gas plays in Western Canada. There might be, he reasoned, "another extensive field like the Viking in the Peace River area which could serve as a supply for Vancouver, some 650 miles to the south."

McMahon also realized it would be the biggest "play" of his life. He would need a great deal of money. Representatives of the British Guinness family interests happened to be in Vancouver, investing in commercial real estate. They urged him to go to London to look for investors and provided him with introductions to the biggest financial and oil concerns in Europe—including Shell Oil and Anglo Iranian. McMahon spent the next five months in England, trying to raise capital; the trip was a social success and a business failure. Although the British geologists all agreed that the Western Canada basin was undoubtedly a major petroleum basin, they had no interest in pursuing it when there were less risky fields in the Middle and Far East. A side trip to meet American oil investors, including Standard Oil of California, also proved fruitless.

Back in Canada, Columbia Oils had started work on the economics and engineering of McMahon's Peace River country project, but in 1935 the B.C. government intervened. Premier Duff Pattullo decided to take over the oil and gas business in the province by the simple expedient of withholding mineral rights from the market and developing them as a state enterprise. Frank McMahon seemed to have reached the dead end of his oil career. Columbia Oils was certainly finished. His financial situation had passed beyond desperate. At thirty-four years of age, with a wife and three young children, he seemed headed for a Depression government work camp as the only means of surviving. At that moment, his brother George later observed, he exhibited what was to become a defining characteristic of his career. He absorbed the setback of Premier Pattullo's exploration moratorium, then, with his cheerful disposition and wry, self-deprecating sense of humour restored, responded by redoubling his efforts to pursue his original goal. Although he had no idea whether the project would ever materialize, he quietly completed the conceptual planning of a gas exploration, production and transportation system from northeast B.C. to the Lower Mainland. "We had engineers draw the first detailed map of the pipeline system as early as 1937," he later recalled.

PUTTING FOUR YEARS of frustration behind him, Frank McMahon took off for Turner Valley. The biggest oil field in the British Empire had just been rejuvenated from its Depression doldrums by Home Oil's discovery of its most prolific oil well. Turner Valley is the only oil field in Canada to have been discovered three times, each occasion setting off a prospecting boom. In 1914, Stewart Herron and his backers brought in Dingman No. 1, touching off the stock frenzy that gave birth to the Alberta Stock Exchange. In 1924, Royalite Oils, owned by Imperial Oil and the Herron family, brought in Royalite No. 4, the well that blew wild in a gigantic fireball, putting Alberta on the world oil map. Now, in 1936, Bobby Brown Jr. and the indefatigable Major Jim Lowery wrote their own page in the history of the field. Turner Valley Royalties No. 1 punched into a previously undetected, deep oil reservoir on the west flank of the field, launching Home Oil as a future major Canadian independent and Brown as the McMahons' lifelong friendly rival.

Frank and George McMahon wanted a piece of the action but seemed shut out. Mineral rights in the area had been leased and re-leased for twenty-two years, and there appeared to be little room for newcomers. Combing through land records, however, Frank came across a 40-acre freehold located just a hundred yards from Home Oil's big strike and owned by a CPR station agent named Moody Shore of Abbotsford, British Columbia. Frank didn't own a car, so he rounded up a fellow named Harry Steveson who did and the two men drove non-stop to Abbotsford, where they gave the surprised Mr. Shore $100 for an option on his land. (McMahon family legend holds that this was the last $100 in Frank's personal bank account.) Next, they sped into Vancouver and scraped together $20,000 from the former Columbia Oils backers. They then drove back to Abbotsford and gave Shore a $20,000 deposit on $200,000 in cash and the promise of shares in the company that would drill the lease.

The McMahon brothers incorporated West Turner Valley Petroleums and raised yet another $200,000 from their Vancouver backers. By the end of the summer of 1937, Frank was back in his coveralls and muddy boots, on the drilling rig floor at Turner Valley. He had one more obstacle to overcome. This was a deep well, below 3000 feet. Short of the target reservoir, he ran out of money again. There was no going back to Vancouver, so he cut a deal with the Royalite Oil Company to finance the last few hundred feet in exchange for a hefty overriding royalty. It was worth it; when the well came in, in the late winter of 1938, it was an elephant, initially producing 3500 barrels of oil per day—the most prolific well in Turner Valley history.

Frank McMahon needed every penny of the revenues, after Royalite took its cut. He paid off Moody Shore, and bought up three more 40-acre leases. West Turner Valley drilled two more successful wells in 1938, each producing 2000 to 3000 barrels a day. Even after the wells settled down to more modest flow rates, the company was banking $1000 to $2000 a day in net income. George sent a young accountant named Pat Bowsher from Vancouver to Turner Valley to help catch up on the record keeping. Bowsher formed a low opinion of Frank's business practices and expressed his feelings bluntly to the startled wildcatter. He went back to Vancouver, certain that his barbed tongue meant he'd never work for the McMahons again. But within a couple of days,

George was on the phone asking him to come in and get the company's books and files organized. Bowsher became one of Frank McMahon's closest friends and his lifelong Chancellor of the Exchequer. George McMahon was the founding member of Frank's inner circle, and Pat Bowsher was the first outsider to be admitted to the sanctum. McMahon was a team player who surrounded himself with the strongest partners he could find. Other men in his position sought sycophants. He preferred to have on his team individuals smarter than he in their area of professional competence. He drove them crazy because he paid no attention to detail, and often they resented cleaning up his messes. But teams Frank McMahon built produced results.

McMahon was getting set to drill West Turner Valley No. 4 in the winter of 1939–40 when one of his Vancouver financiers, George Martin, brought in a merger proposal. He controlled a company called British Pacific Oils with few assets but a well-heeled group of shareholders. A merger would tickle more investment capital out of that shareholder base, he promised. On January 21, 1939, West Turner Valley and British Pacific Oils merged to create Pacific Petroleums, and the head office was transferred to Calgary. Frank, George and their families moved with it.

The outbreak of war in September 1939 cooled the drilling frenzy in Turner Valley. However, the area's production became an important source of aviation fuel for the British Commonwealth Air Training Plan, the multinational program that trained new fighter and bomber pilots in Alberta because its terrain closely mimicked Europe's battle conditions. With the invocation of the War Measures Act, the federal government appointed an Oil Controller and created a Crown corporation called Wartime Oils that funded one-third of the cost of successful oil wells. Pacific Petroleums drilled two wells at Turner Valley using Wartime Oils' funds, as well as exploring on its own for heavy oil and natural gas in the small fields of central and eastern Alberta that were the only successful discoveries outside Turner Valley. With the Canadian oil industry in hiatus, the McMahons hedged their bet on Pacific Petroleums. George stayed in management and together he and Frank held what amounted to the controlling equity interest. However, Frank soon resigned and went back to the diamond drilling contract business. In his extra hours, he established a string of sister companies to Pacific Petroleums that incurred higher risks on wildcat drilling than Pacific, with its growing shareholder base, was able to absorb.

ONE DEVELOPMENT HAD a longer-term significance. Frank McMahon discovered two small gas pools near the villages of Taber and Princess in southeast Alberta. The fields were of no consequence to his financial situation but sparked an important event for the future of the oil sands. Frank wanted to sell the Princess field and finding no Canadian buyer, he looked south to American investors—a practice that he would follow increasingly as the war drew to a close. This time he fell back on his Whitmore College connections. John Howard Pew, the patriarch and controlling shareholder of the Sun Oil Company, was also an active churchman and devout Christian. He gave his extra suits to a young evangelist named Billy Graham and his money to Whitmore College. McMahon went to see Pew and was so persuasive that Pew bought the Princess leases and production, thus ushering Sun Oil into Canada.

After the war, under Pew's direct leadership, Sun acquired leases in the Athabasca area near Fort McMurray. It financed the world's first commercial oil sands mine and synthetic crude oil extraction unit, called the Great Canadian Oil Sands Plant. The project used technology developed by Karl Clark at the Alberta Research Council, and economics developed by his former graduate student Sid Blair, who was president of Canadian Bechtel when Pew engaged the firm to design and engineer the plant. Construction commenced in 1964 and production started up in 1967.

In parallel with these events, Pacific Petroleums acquired a spread of oil sands leases after Premier Ernest Manning's key energy bureaucrat, Deputy Minister Hubert Somerville, courted the McMahons as oil sands developers. Technically, a single company could own only 10,000 acres of oil sands leases. George and Frank sat by George's trusty Underwood for a whole day composing five hundred names for five hundred subsidiaries to hold 10,000 acres apiece. This early foray into oil sands sidetracked McMahon's time and energy only briefly. He soon realized that oil sands development would grow at a slower pace than conventional activity, so he returned to more aggressive opportunities. But he held onto the oil sands interests and in 1964 Pacific Petroleums joined what became the Syncrude Canada consortium, though its plant at Mildred Lake wasn't completed until 1974. The interests that George McMahon acquired with Hubert Somerville's assistance in 1948 passed to Petro-Canada after the Crown company purchased Pacific in 1979.

In 1947 and 1948, however, as John Howard Pew was undertaking his first studies of the oil sands and starting the process of acquiring leases, Frank McMahon was chasing the play that finally made him rich. He travelled from Burnt Timber in western Alberta to Cat Creek in northern Montana, leasing land for new drilling prospects for Pacific Petroleums. The hottest action, however, was in central Alberta. From the first rumours, in January 1947, that Imperial had struck oil in a Devonian reef at Leduc until the company brought in its well, in a stage-managed publicity stunt on February 13, land men for competing oil companies swept through central Alberta looking for oil leases. A tiny outfit named Atlantic Oil Co. purchased a freehold quarter-section for $200,000 and drilled two uninspiring wells on it. Meanwhile, Pacific Petroleums came up empty-handed in its search for drilling locations, until early in 1948, when Frank decided to buy a controlling interest in Atlantic.

Frank purchased the Atlantic equity and pumped every cent he and George could raise into the new venture. Within a few weeks of the acquisition, McMahon had a tiger by the tail. Atlantic No. 3 in the Leduc field blew wild and remained out of control for six months, spewing 1.25 million barrels of oil into a network of hastily constructed earthen berms. The flow from the well was so powerful that it opened fissures in the ground hundreds of yards from the wellhead. "We were the only company in Canadian history to test a well through a 40-acre choke," Frank quipped. Well-control experts poured millions of gallons of water and a crazy assortment of junk down the well bore to staunch the flow of oil: bundles of wire rope, golf balls, grain seeds, horse manure, sawdust, waste cotton and ten thousand bags of cement—the biggest single order to that date for Alberta's concrete industry. Nothing would plug the ranging torrent. Finally, it caught fire and two relief wells were completed to draw off the flow of oil, enabling the crew to shut in Atlantic No. 3. It never produced again. In retrospect, Frank enjoyed telling the story of an embarrassing promise he'd made to John Rebus, the farmer who owned the land on which Atlantic No. 3 was drilled. When he settled the deal for surface access, which included cash and a washing machine for Mrs. Rebus, Frank promised Mr. Rebus, "When this well is drilled, you'll never know we've been here!"

After clean-up costs were met and damage claims paid covering the share of oil drawn off neighbouring wells, Atlantic Oil netted $1.25 million, or $1 a barrel from oil recovered during the blowout. Production was reinstated in 1949 on Atlantic No. 1 and No. 2, and two more wells drilled in early 1950. Finally, the McMahon brothers had their company-maker. They merged their Atlantic equity into Pacific Petroleums, and Frank in 1948 resumed his duties as Pacific's managing director. A few months later, after an unexpected and bitter dust-up with other founding shareholders, he was appointed president, managing director and chief executive officer; thereafter, he and George effectively controlled the company, though they owned less than 30 percent of its equity.

Preoccupied with the wild well, Pacific Petroleums missed the early window on exploration opportunities that followed the Leduc find. In September 1948, shortly after Atlantic No. 3 went quiet, Imperial Oil discovered a second huge Devonian oil field at Redwater, west of Edmonton. The McMahons were free to strengthen Pacific's hand in the exploration rush that consumed Alberta. With a new American partner, Sunray Oil of Tulsa, Oklahoma, George acquired new drilling acreage on the edge of the Redwater play, while Frank concentrated on finishing the effort to quell the rogue well at Leduc. In one Crown auction for leases, George paid a record $1 million for the exploration rights on a section of land (640 acres). This milestone was indicative of the intensity of the oil boom, and of the kind of money companies spent on it. Within eighteen months, the Pacific-Sunray partnership drilled forty successful oil wells.

In spite of the controversy surrounding Atlantic No. 3, the McMahon brothers had reached the top of the oil patch pecking order. That much became clear when the fabled California oil billionaire John Paul Getty took a 40 percent interest in their next venture, Bear Oil Co. With Pacific Petroleums as an equal 40 percent partner and Sunray holding the remaining 20 percent, Bear Oil assembled four million acres of leases and hired retired Imperial Oil geologist Ted Link to pick its drilling targets. The legendary Link was the most successful petroleum explorer in Western Canadian history. He had directed the discovery of Leduc and, in 1921, the Norman Wells oil field in the Mackenzie Valley. With these two finds, Link solved the mystery of oil occurrence in Devonian-era geological formations that provided most of the oil produced in Western Canada from 1950 to the end of the century, excluding the oil sands. But Link drilled his share of dry holes, too; the magic touch failed Frank McMahon and his partners. Bear Oil spent $5 million on a two-year seismic survey and drilling program and found nothing. Getty moved over to Saskatchewan and spent another $8 million on wildcatting before pulling out of Canada and devoting his capital to the concessions in the Middle East that made him famous. The McMahons turned their attention west—back to British Columbia's Peace River country.

3

The Peace Country Project

All through the western world there is a good deal of talk about newly discovered resources in Canada, oil, iron, copper, nickel, titanium, uranium. People talk about petrochemicals and wood chemistry. They talk about all the industries that go with waterpower.

DR. W. A. MACKINTOSH

The Second World War propelled Canada out of a rural, agrarian economic slumber into the mainstream of North America's industrial economy. Not only had the Allied war machine soaked up Canada's food production and its oil, iron and mineral resources, Canadian industrial plants had been retooled and expanded to build tanks, trucks and aircraft. Capital from the United States poured into its investment markets. In Western Canada, expanding oil and gas production, which was little more than a cottage industry in 1939, was now at the top of the economic agenda. Before the war, Canada produced only 10 percent of its own oil consumption and its oil reserves amounted to less than a year of equivalent consumption. Five years later, Alberta produced enough oil to supply everything the Prairies needed, and Canada had the equivalent of twenty years' consumption in the ground. Frank and George McMahon were one of the leading reasons the situation had changed so emphatically.

In 1945, Frank formed a consortium of Pacific Petroleums and an investor group of he, his brother George, department-store magnate Vic Spencer and financiers George Martin and Norm Whittall. The partnership applied to the British Columbia government for a permit to assemble a 700-mile right-of-way from Fort St. John to Vancouver, and to acquire a 1.25-million-acre parcel of petroleum leases in the Pouce Coupe area. The initiative was buoyed by the euphoria of the war's end but was ahead of its time. The province still wanted to develop Peace River gas as a Crown enterprise and launched its own two-year drilling program. The venture failed and became a political powder keg.

In 1947, the British Columbia government lifted its moratorium on private investment after sinking $1 million into a failed exploration well at aptly named Commotion Creek. The well could never be completed after the drill bit and column got stuck in the hole and couldn't be retrieved. McMahon, who followed the project closely, stepped in to bail out the drilling contractor when the debacle drove it to the verge of bankruptcy. The public anger at the waste of government funds ended the dream of a state petroleum monopoly. The provincial government offered petroleum leases to private investors and Frank McMahon camped out overnight in front of the title office in Victoria to file for the first three petroleum prospecting leases that Pacific Petroleums intended to drill. He then sent rigs into British Columbia to drill the first raw wildcats of the gas play.

In November 1951, Pacific Petroleums completed Fort St. John No. 1, the first commercial gas find in the region. That winter, Pacific Petroleums had six rigs drilling constantly in the area, and it was months before the first competitor—Phillips Petroleum of Oklahoma—showed up to challenge them. McMahon assembled another consortium, of Pacific Petroleums, Sunray Oil Co., Peace River Natural Gas (a Pacific Petroleums subsidiary), Hudson's Bay Oil and Gas, and Union Oil of California. The consortium jointly held 777,830 acres. With this area added to the 545,131 acres controlled through Peace River Natural Gas, Pacific Petroleums' leadership of the consortium gave it dominance over oil and gas exploration in the province.

McMahon had meanwhile dusted off the 1937 engineering map of a pipeline from Fort St. John to Vancouver. On April 30, 1949, Canada's Parliament voted to incorporate five pipeline companies including the Pacific Petroleums–sponsored Westcoast Transmission Company. Getting a start on pipeline planning more than two years before the first discovery gave Westcoast a lead that no competitor was able to close.

Less than a week after Westcoast's incorporation, on May 6, McMahon added another key member to his inner circle, a legal whiz whom he'd gotten to know and admire through the Bear Oil project. Douglas Peter McDonald, K.C., was a senior Calgary oil and gas lawyer with deep contacts in government and lengthy experience in the development of postwar regulatory agencies. McDonald, who insisted on being called "D. P.," was the son of a mine carpenter and union leader who became the mayor of Rossland, a tiny town in British Columbia's West Kootenays. A high school baseball, hockey and lacrosse star, D. P. learned to play football at the University of Alberta, where he studied law, and went on to play semi-pro football and hockey as a young lawyer learning the oil and gas business representing Turner Valley wildcatters.

D. P. did the legal work for McMahon's Pacific Petroleums and the dealings with J. H. Pew. He moved further into the McMahon orbit at Bear Oil, structuring the partnership and, as joint manager, overseeing the expenditure of funds on behalf of the silent partners. He had the McMahons' and Pat Bowsher's respect as a deal-making, not a deal-breaking, lawyer. After winding down the activities of the company, McDonald served as counsel to the Dinning Royal Commission in Alberta. The commission was created by Premier Ernest Manning to determine at what point sufficient natural gas would be discovered in Alberta for the province's future internal needs, so that removal permits could be granted to ship surplus gas out of the province on proposed pipelines. The commission finished up in 1949—recommending no exports until more gas was found—at the same time McMahon was organizing Westcoast Transmission. A week after the incorporation cleared Parliament, McMahon hired D. P. McDonald as general counsel and retained him as his personal lawyer. This dual role, his solid business acumen and his undisputed first rank among an emerging breed of regulatory lawyers cemented McDonald's position as the most powerful corporate counsel in the oil patch. McDonald remained with Westcoast for twenty-one years, then became the chairman of Westcoast Petroleum before retiring in 1972.

Frank McMahon completed the inner circle of executives who navigated Westcoast Transmission through its formative years by recruiting MIT-trained engineer Dr. Charles Hetherington. Charlie Hetherington, affectionately known as "Dr. Charles" by many Westcoast Transmission staff, was the son of a Norman, Oklahoma, physician who studied petroleum engineering at the University of Oklahoma, then fled the state as the Depression turned the Oklahoma oil patch into a ghost town. He worked for Shell Oil and got his doctorate. During the war, he joined the U.S. National Defense Research Committee, then hired on in 1946 with the world's top gas pipeline engineering firm, Ford, Bacon & Davis of New York. FBD had been involved in pipelines in Western Canada since it designed and supervised construction of the first Alberta gas line in 1912, a 16-inch trunk running 170 miles from the Bow Island gas field to Calgary.

Hetherington turned up at McMahon's Calgary office in the winter of 1948 with two other senior Ford, Bacon & Davis engineers to design and plan construction of Westcoast Transmission's line. He stayed to become McMahon's chief assistant and the company's spokesman on engineering and technical matters through the long regulatory struggle ahead. Hetherington became much more than Westcoast's chief engineer and the principal architect of the pipeline and head foreman during construction. He shuttled back and forth between Pacific Petroleums and Westcoast for twenty years, from 1950 to 1970, developing projects, and acting as McMahon's personal troubleshooter and, when necessary, hatchetman. Titles didn't matter a great deal to Hetherington: his highest rank was vice-president of corporate development. But he made a fortune on shares in both Pacific Petroleums and Westcoast Transmission, and enough to do McMahon's bidding with complete confidence. D. P. McDonald became the company's anchor; Charlie Hetherington was its sail.

Two other men deserve mention for their position on the inner circle's fringe. Len Youell was hired by Pat Bowsher in 1945 and worked in the executive office in Calgary and later in Vancouver as an executive aide for the two McMahons and Bowsher. When Phillips Petroleum gained control in 1963, he stayed seven more years as corporate secretary, retiring in 1970. Another engineer, Peter Kutney, joined Westcoast Transmission when the line went into service as gas supply manager; he built the cornerstone of the company's marketing system. He stayed with Westcoast until 1971 and went on to found a successful junior company called Coseka Resources, which thrived during the halcyon days of the oil boom. Kutney, Hetherington and McDonald became the Three Musketeers of the company's operations.

While Pacific Petroleums searched for its first Peace River gas field through the drilling seasons of 1949 and 1950, Westcoast took the initial practical step to build an integrated natural gas system linking the wells of the Peace country to the furnaces of Vancouver. The company completed a small, 6-inch-diameter pipeline link from Pouce Coupe, Alberta, over the border to Dawson Creek. The line followed the hard-won approval of the Alberta government in 1950 to ship natural gas out of that province. It launched British Columbia's first home delivery of natural gas, in the Dawson Creek area.

Frank and George McMahon hoped to incorporate northwest Alberta gas production into their future pipeline network. It didn't work out that way: the line never functioned reliably and within six years the great TransCanada PipeLines project drew Alberta gas eastward. However, the launch of the Peace River line at Dawson Creek on October 31, 1950, served notice that Westcoast and its ambitions were not going to fade away. "It was the biggest ceremony for the smallest inter-provincial line in the history of Canadian pipelines," George McMahon said of the whoop-up, which featured the ignition of a flare. When the pencil of flame shot into the Halloween night air, Frank's memory flashed back twenty-five years to that night on the Pouce Coupe River with the sky illuminated by the wild-burning gas well. Having spent two decades building a successful natural gas production company, McMahon was now on the threshold of a ten-year battle to attach an international pipeline to it.

THROUGH THE FIRST HALF of the 1950s, Pacific Petroleums invested $30 million to develop gas production in northeast British Columbia, while Frank McMahon completed the detailed engineering for his project and battered at the regulatory doors in Canada and the United States to build the Westcoast Transmission line. Thanks in part to the advocacy of his old friend George Hume, the postwar director of the Geological Survey of Canada, by 1952 Canadian officials in Ottawa, Edmonton and Victoria had accepted the idea of Canadian gas exports to the U.S. In April that year, McMahon applied to the United States Federal Power Commission (FPC) for what he thought would be a routine approval to import Canadian gas into the Pacific Northwest states.

He expected to complete construction and start moving gas by the end of 1954, but he had not counted on stiff American competition: the Pentagon challenged the idea of dependency on Canadian energy, and a consortium of four American firms—Pacific Northwest Pipeline, El Paso Natural Gas, Colorado Interstate and Pacific Gas and Electric—wanted to supply McMahon's prospective market from fields in Texas and New Mexico. McMahon's competitors tied up his proposal during two years of FPC hearings. He was sitting in his barber's chair at the Hotel Pierre in New York on June 18, 1954, when he received a phone call that the FPC had rejected the Canadian project and determined that Pacific Northwest Pipeline Corp. should have the prize. Pacific Northwest had its own problem: it had a certificate to build a pipeline into northern California, Oregon and Washington from the east, but it didn't have a ready gas supply. McMahon brooded for less than a day. Then he gathered his team to find a way to fit Peace River gas into the American market by finding common ground between Westcoast Transmission's proposal and Pacific Northwest's lock on the FPC.

He approached Pacific Northwest chief executive Roy Fish with a proposal to merge the two projects. Westcoast would bring Peace River gas to the border and interconnect with Pacific Northwest at the Sumas border crossing in the Fraser Valley near Vancouver. In a side development, a B.C. utility called Inland Natural Gas would set up service to the province's southern interior communities. It took a year, but Fish was captive to the gas reserves Pacific Petroleums held in northeast British Columbia, and in late 1954, Westcoast, Pacific Northwest and El Paso Natural Gas signed agreements locking up a deal.

On January 1, 1955, in anticipation of regulatory approval, McMahon appointed Canadian Bechtel Limited the project engineer and commissioned Ford, Bacon & Davis to design the compressor stations. The plan was to construct a 30-inch, high-pressure line with three compressors in 1956 and 1957. At start-up, the line would carry 350,000 cubic feet of natural gas per day. The addition of four compressors over the first years of operation would raise capacity to 510,000 cubic feet per day. Canadian Bechtel started work on February 1. Through the year, McMahon aggressively drove the project along, taking U.S. Federal Power Commission approval for granted, so that when the go-ahead came on November 25, Westcoast had spent $19 million on design, engineering and initial construction. Exposing $19 million of work before approval, however, wasn't the biggest risk McMahon and his various financial backers had taken.

The fact of the matter was that he hadn't actually drilled up and connected the gas reserves as he claimed, and that claim had been accepted by the regulators and his future customers as justification for the project. Now Pacific Petroleums launched a seventy-well drilling program to make good on McMahon's commitments. It was a harrowing time. Charles Hetherington loved to tell the story of a California Public Utilities Commission meeting at which he and Frank McMahon were briefing some of the state's technical staff. One American engineer asked McMahon the fairly routine question, "What is the BTU value of the gas in your reserves?" McMahon couldn't answer because no gas had yet been produced to be analyzed—a fact that would have likely killed the project had American regulators realized it. Stumped, McMahon said, "I think I'll let my expert, Dr. Hetherington, answer that question." Never at a loss for words or charm, the outgoing Hetherington smiled calmly and replied, "I don't have the BTU value of our gas at my fingertips, but I know that when you light it, it burns!"

As 1955 ended, the long wait was over, and the pipeline construction was underway. No one, least of all McMahon, doubted that the gas fields would prove up in time to fill the line.

THROUGH ALL HIS YEARS bringing the Westcoast Transmission dream to fruition, Frank McMahon's private life had run on a parallel but separate track. In 1935, he had been a penniless wildcatter not certain how he could provide for his family. After Atlantic No. 3 blew wild, Frank and Isobel McMahon became the millionaires they had wanted to be since they'd wed in the summer of 1928. In Vancouver they maintained a fashionable home on Marine Drive and entertained lavishly a circle of associates including Frank Ross, Norm Whittall and John McMahon. Frank had a life outside the oil business, and it usually started with a four-martini lunch after he had worked at his office for six to eight hours, since dawn. In Calgary, the McMahons enjoyed a growing circle of friends including Home Oil's Bobby Brown, philanthropist Eric L. Harvie, Conservative MP and oil commentator Carl Nickle, Turner Valley kingpin Bill Herron and newly-rich independent wildcatters Charles Lee and Sam Nickle. Publisher, rancher and oil investor Max Bell had drawn Frank into the horsy set and, in spite of his Presbyterian piety, he threw great parties. (Once when Max and Frank were flying in a Pacific Petroleums charter, an engine conked out during a fierce thundershower. In a flash, Max was on his knees praying for salvation and pleading with Frank to join him. The agnostic McMahon patted Bell on the shoulder and carried on pawing through a briefcase full of undigested paperwork.)

As a signet to their new petroleum wealth, McMahon and some of his friends founded Alberta Distillers Ltd. in 1950. While they waited for their first batches of rye whisky to age quietly in their barrels, they decided to make vodka, which requires no maturing. Their timing was impeccable: the New York fashion for vodka martinis, Moscow Mules and Bloody Marys had started and McMahon made another small fortune exporting Alberta vodka to the United States. His business interests expanded and extended into copper mining, manufacturing and real estate.

The early years of his wealth presented Frank with two tragedies. His son, Frank Jr., was killed in an automobile accident in 1952. His mother, Stella, was murdered. He drew a veil over his sorrows. When, several years later, he buried his brother John, he told Pat Bowsher, "Life was never guaranteed to be fun. You have to grab it and shake it to survive."

By 1954, Isobel McMahon felt the distance between her and Frank had become intolerable. He had become a patrician figure in his fine suits and perfect coiffure, at home in the cosmopolitan milieu of London and New York. With his free-spending ways and his sense of high style, he was a very appealing man, and rumours abounded. Stories of his extra-marital conquests got back to Isobel. His pals nicknamed him "Big Inch," not only because of his pipeline. The marriage survived until their daughter Marion was married in June 1954. It was an extravaganza, with guests flown in from across North America by McMahon's beloved Lougheed Lodestar jet. The McMahons took over the best suites in Calgary's prestigious Palliser Hotel, the unofficial petroleum club of the day. The morning nuptials were celebrated in the Cathedral Church of the Redeemer, amid a virtual forest of white peonies, carnations and gladioli, arranged by a Hollywood set designer. Marion walked to the altar across a bed of petals, followed by eleven attendants, while a thirty-voice choir sang the wedding anthem. Back at the hotel, the wedding guests celebrated over a champagne and chicken brunch, then stayed for the weekend to party and gossip while the bride and groom left for an exotic honeymoon. The wedding set McMahon back $20,000—then the equivalent of a year's pay for a top executive. (It would have cost more but Mantovani's orchestra had another booking in Calgary and had no need to take up Frank's offer to fly it in from London.)

A year later, Frank and Isobel divorced. In 1956, Frank was remarried—too quickly for proper Calgary society—to New York fashion designer, teen magazine columnist and socialite Elizabeth Bettes. Although they retained a large home with swimming pool in the Calgary, the couple's primary residence was a luxurious Park Avenue apartment overlooking New York's Central Park. They also maintained permanent suites in Ottawa's Chateau Laurier Hotel, Vancouver's Bayshore Inn, the Beverly Hills Wilshire and San Francisco's Mark Hopkins. Later in life they added a Palm Beach "cottage" that could seat one hundred for dinner and a French colonial home in Vancouver.

Betty McMahon taught Frank how to enjoy his money. She introduced him to a Hollywood crowd far racier than the gang that Bing Crosby—good Catholic and husband of one wife—hung out with. She brought him together with actress Rosalind Russell, who had just achieved fame starring in the Broadway production of *Auntie Mame*. Russell's husband, producer Fred Brisson, seeking a meaningful career to balance his wife's success, was backing the Broadway lyricist and composer team of Richard Adler and Jerry Ross, who had just written a musical called *The Pajama Game*. By coincidence, McMahon had just read and enjoyed the book, *Seven and a Half Cents*, on which the story was based, so he gave Brisson U.S. $10,000 to top up his funding. It wasn't enough, and Frank threw in another U.S. $70,000 before the show went on its road tryouts.

He had pretty much written off the investment as an interesting experience and an expensive but effective membership fee to the exclusive showbiz crowd in Manhattan, when *The Pajama Game* opened in the St. James Theater. It had received rave reviews on the road, which McMahon hadn't noticed, so he was shocked when it turned out to be a box office smash hit. His expected financial loss had become a horrendous tax problem. He reinvested the profits in another Brisson production to avoid taxes, but the Adler and Ross musical *Damn Yankees* was so successful that it escalated Frank into the 70 percent tax bracket. (His estate still draws earnings from his rights.)

Betty McMahon also gave Frank his entree into the top ranks of American horse racing, a passion he had shared as a partner of Max Bell. Three of his horses were track superstars: Diablo, Majestic Prince and Irish Dancer, owned in partnership with Max Bell until Bell sold his share to Bing Crosby. Diablo, the horse Frank loved best, was his entry into the Kentucky Derby. As an unknown newcomer to the ranks of the Triple Crown, McMahon was under pressure to sell the horse to a more established stable before the big race. He refused, and in a preliminary run in San Francisco another horse collided with Diablo, injuring him so severely he couldn't run again. McMahon was suspicious ever after about the circumstances of the accident, and remained bitter about it for the rest of his life—though he never brought himself to make a public accusation that Diablo had been targeted.

In 1969, Frank McMahon capped his racing career with one more try for the Triple Crown. His horse, Majestic Prince, purchased in Kentucky for U.S. $200,000 as a yearling in 1967, won seven races in a row in his first two years under McMahon's colours. In his final appearance before the three great races, Majestic Prince captured the coveted Santa Anita Derby. The 1969 Triple Crown series was a battle between Majestic Prince and Arts and Letters, owned by Gulf Oil heir Paul Mellon. At Churchill Downs, Majestic Prince won the Kentucky Derby. Frank's guests in his private box that day included American president Richard Nixon and a successor, Ronald Reagan. The Prince beat Arts and Letters again to win the Preakness cup. Then, in one of racing's greatest controversies, jockey and trainer Johnny Longden scratched Majestic Prince from the one-hundredth running of the Belmont Stakes, the third jewel in the crown. A frustrated and angry McMahon overruled the trainer and took his champion to New York. In spite of enormous fan support, Majestic Prince finished his final race five lengths back of Arts and Letters, and he was put out to stud. While his fees provided McMahon with another lucrative tax problem, the disappointed owner left the racing circuit, one arena of endeavour in which he had not won it all.

George McMahon had a very different lifestyle. He had settled into the Calgary community and become a popular figure around the Petroleum Club and the United Way. In 1956, he was given a coveted directorship on the board of the Calgary Stampeder professional football club. George had been a good Canadian university-level player, loved the sport and was well known in football circles, so he had enough leverage to persuade the team's insiders to include his brother Frank on the board.

Frank stayed with the team only six years. However, he played a key role in 1958, with the NHL Hockey Hall of Fame's Red Dutton and youthful lawyer Pat Mahoney (later a Trudeau-era Liberal MP and federal court judge), in creating the Canadian Football League out of the eastern Big Four and western professional leagues. Frank led the three Calgary representatives into the owners' meeting in Winnipeg that year and forcefully made the case for the merger. Mixing charm for his supporters and intimidation for the wavering delegates, he kept the discussion going until he could count on the unanimous vote.

George was president of the Calgary Stampeders in 1960, when the club badly needed a new playing field because it had outgrown the historic but shabby downtown Mewata Stadium. He persuaded Frank that they should contribute $300,000 to build a twenty-thousand-seat stadium and field house on University of Calgary endowment land. The football club would be the anchor tenant and would manage the field under a contract with the non-profit McMahon Stadium Society, while the university's sports program would be given free access to the facility. In the summer of 1960, George oversaw completion of the stadium in 103 days from the first shovel in the ground to the last daub of paint on the seats. The football team launched its first season there on August 15, 1960. Since then, the expanded field has hosted two Grey Cups, and the opening and closing ceremonies of the 1988 Winter Olympic Games.

Frank moved easily among the business cultures of Calgary, Vancouver, New York, Los Angeles and London. His gypsy lifestyle infuriated his staff and business partners—he became notorious for missing meetings because he was at the opposite end of the country playing poker or at one of his favourite racetracks. Most months, he slept more often aboard his executive jet than in any of his residences. After his remarriage in 1956, he and Betty lived the life of American corporate royalty, moving among their various homes and offices. He hovered around the Vancouver office during construction of the Westcoast Transmission pipeline through 1956 and 1957, making frequent trips to inspect the work. But those closest to him could see that completion of the line would be a turning point in his life, and that he would be an absentee boss once the line was built.

4

The Pipeliners

*I arrived in Fort St. John in February of 1955 to be a right-of-way
agent. About the time I got there, there was just a real
cold snap. They had tapped a well right outside town and everyone
was dependent on gas for cooking and heating. That month
it got so cold—minus 40 Fahrenheit—the well froze off. The steam
pipes froze in the hotel; there was no heat in many homes.*

BERNIE GUICHON

The construction of the Westcoast Transmission pipeline was the adventure of a lifetime. The men and women who lived it understood that in a personal way. They also understood that this was more than an energy project; it was an event that would change and thus define British Columbia, and in combination with other such developments of the decade, contribute to the new, postwar, industrial Canada.

Bernie Guichon, a right-of-way agent during construction and later manager of lands, right-of-way and environment, carried the vivid memories with him throughout his long career and into retirement:

> The only place you could get warm was in your car. Everybody was leaving their cars running and sleeping in their cars. The man who ran the Chinese restaurant had a wood stove and a wood heater that

he fired up; there was hot food and a warm place to eat, but everything else was frozen up.

It was my first time in Fort St. John and the first time I'd been over the Peace River Bridge. There was a saying that if you go over the bridge once, you'll always come back. The way the weather was, I didn't know whether I would ever come back. But I did.

Guichon had joined a Canadian elite. They were the pipeliners. tough, wilderness-wise, travelled men with sun-weathered faces and callused hands blackened by welding rods. They came from the six corners of the oil patch to lay pipe and build compressor stations in the hills and valleys of British Columbia's northern interior. They trickled in during the first months of 1955; the following spring, the trickle turned into a flood, bringing with it the varied accents of Texas and Louisiana, Wyoming and Pennsylvania, Ontario and Alberta. The pipeliners' voices blended with the soft-spoken cadences of the Indians, Metis, local trappers and homesteaders whose peaceful isolation they disturbed, and whom they occasionally recruited to their renegade band.

Harry Dickie was one of those trappers lured out of the bush for pipeline pay. He had augmented his fur income by doing janitorial work in a Fort Nelson school; when the pipeline came, one of the foremen heard of his reliability and hired him to clean up around the compressor stations and workshops, and haul garbage. Said Dickie many years later, "It was better pay than the school and less arduous work. They treated their men well."

After two years, Dickie went back to trapping, but the pipeline had changed his life, opening up his horizons to the world of the 1960s in a way that even Army life hadn't. Although he had completed only grade nine himself, one lasting influence of the pipeline was that he "made sure [his] four children all completed university." At eighty-six years of age, when interviewed for this book, he still trapped, "not for the money but just to enjoy life." Along his traplines in the silence of winter he carries the memory of the good men he knew on the pipeline, "too many of them gone now, retired or passed on." The world of the Peace Country was forever changed by the advent of natural gas and oil. The contingent of pipeliners grew, by summer 1955, to a force 2200 strong. They were the infantry grunts of North America's dizzying economic expansion. They had laid down the guns of European and Pacific battles and picked up the tools to build the highways, dams, cities and airfields of postwar prosperity. Ancient scripture had never rung so true; for a decade or more, an entire continent beat its swords into ploughshares.

Time and again, the pipeliners had been asked to cross the toughest terrain that geography could offer. Now the drama of a wild land yielding to work gangs was re-enacted on a stage of pipeline construction sheds 650 miles long. The unslaked thirst for oil and natural gas drove these construction warriors ever northward, deeper into Canada's endless vault of resources.

Pipeliners had a special reputation for overcoming nature's obstacles. They had been schooled in laying pipe across the toughest environments, from the deserts of Arabia to the snake-infested bayous of Louisiana, across Appalachian summits and through the untamed Rocky Mountain passes. Arriving in the British Columbia Interior, they surveyed a land that for centuries had daunted the most intrepid fur traders and gold prospectors. With optimism bordering on bravado, they called it "good pipelining country," deeming it no more difficult to cross than terrain they had carved through elsewhere. Pipeliners, characteristically, talked big. But when they stumbled out of the crew buses at first light to fire up the equipment, they always found a way to deliver the goods—even if that meant finding a way to lay pipe down a seemingly vertical slope, or submerging it in an apparently bottomless swamp.

In the 1950s tens of thousands of miles of new pipe were buried between dozens of oil and gas fields and hundreds of cities and towns. To keep up with the scope and pace of construction, the pipeliners lived like nomads with caravans of exotic-looking heavy equipment: bulldozers, side boom tractors, ditchers, backfillers, pipe wrappers and welding machines.

Along the line, few men bothered with the hard hats, safety glasses, steel-toed boots and thick gloves that became mandatory in later years.

They worked barehanded as often as not, and they wore farmers' rubber boots to keep their feet dry, covering their heads with battered cowboy hats, dusty railwaymen's caps and grease-stained fedoras. They smoked hand-rolled cigarettes lit with raspy Zippos, chewed tobacco and drank rye whisky or navy rum.

It was not considered women's work, except for cooking or housekeeping in the camps that accommodated the men on remote jobs. Mostly, the men's wives and sweethearts travelled with them like the camp followers of some medieval army. While the single men stayed in hotels, the families shared encampments of vacation trailers migrating as part of a mobile rearguard to the construction spreads burrowing into near-empty hinterlands.

Some of the pipeliners had colourful names to fit their eccentric natures, such as Foxy Wolf and Shorty Chandler. Other names reflected the cultural stew produced on a continent of immigrants: Van Buuren, Thorkalson, Thorpe, Peterson, Irwin, Harrold. Among these men was Bernie Guichon, and in coming to the interior of British Columbia from Manitoba, through endless Prairie blizzards of February 1955 to help build the Westcoast Transmission main line, Bernie was returning home.

He was the grandson of a French adventurer and gold prospector who had arrived in western North America in about 1860 at the age of twenty-one and travelled to B.C.'s Barkerville gold rush. He learned English and got into the cattle business, and with his sons he created the Guichon Cattle Company, which had several major ranches on the Interior grazing lands. During the 1940s, the company boasted 4000 head of cattle, 500 horses and 1500 to 2000 sheep. Bernie's father, Lawrence, was the ranch manager, but Bernie joined the Royal Canadian Navy during the Second World War, and it seemed he would never return to the ranch. He took an agricultural engineering degree at the University of British Columbia, specializing in soil. This led him to an early career building dams and other works for irrigation projects, then for Canadian Bechtel on the first TransCanada PipeLines lateral, built from Michigan into Ontario to create a gas market ahead of the completion of the main line from Western Canada.

When Bechtel built the Trans Mountain oil pipeline from Edmonton to Vancouver in 1951–52, the line cut across an old Guichon ranch. Intrigued by the pipeline project, Bernie asked to be assigned to it, and soon found that he had fallen in love with the pipeliners' way of life. When the Trans Mountain line was completed in 1954, he worked in Ontario and Saskatchewan on the preliminary engineering for the TransCanada PipeLines main line. Guichon remembers how he first became involved with the Westcoast pipeline project:

> We had parked our trailer for the winter at Indian Head, and it was frozen in solid. We'd just had the first real snowstorm of winter, and I got stuck in a snowdrift or two and didn't get home until 10 p.m. My wife said the project manager in Winnipeg, Bill Ralston was trying to get hold of me and to phone him no matter what time.
>
> I called him and he said, "Ernie, Guard, the Bechtel land manager, wants you out in Vancouver. It looks like the Westcoast job is going ahead, but this TransCanada job is going to be slowed down politically." I said, "Jesus, I'm frozen in here and I'm nice and warm, and I don't know about moving at 35 below and it's supposed to go to 40 below tomorrow" Bill said to me, "If I were you, I'd get my ass in gear and get moving."

So Bernie and his wife dug out their trailer, hitched it up and started driving for the West Coast. They couldn't drive it beyond the Okanagan Valley because the highways to the coast were blocked with snow, but by mid-February, Guichon was in Vancouver at Westcoast's head office, working to locate the new line and its facilities, figuring out how to acquire the land rights and local permits needed to build the line. He recalled what the work involved:

> In the boonies, nobody new what a really big gas pipeline was, so I was going around explaining it to the local fish and wildlife officer, the owners of ranches, the mayors, the senior citizens who held land. We avoided a lot of problems by listening to a lot of people. I had the

advantage that a lot of them knew my name and knew my dad. On the other hand, I had a reputation to keep up. The Guichons had a pretty good name in British Columbia, and I had to look after that, to shave in the morning and look at myself and remember I was a Guichon and I had to tell these people the truth.

Bernie Guichon stayed with Bechtel after construction, even though many of his colleagues signed on with Westcoast for the permanent operation. Then, Bechtel sent him to Calgary to plan the Alberta Natural Gas line from Crowsnest Pass to California, so he and his wife hitched up the trailer and moved on. He came back to British Columbia, and Westcoast Transmission, to manage lands and rights-of-way when Bechtel built the Westpac oil line from Fort St. John to Kamloops. That pipeline became the founding asset of Westcoast Petroleum; meanwhile, in 1964, Bernie had been transferred to Westcoast's payroll to manage its rights-of-way.

> Frank McMahon fostered an industry where there was none before, and in so doing he helped create jobs, homes, income and a good life in a new land. The pipeline was completed four decades ago. In the intervening years, the oil and gas industry in the Peace, like the industry across the nation, has experienced both peaks and valleys. But throughout it all, oil and gas has always held out a promise for the future for residents of the Peace country.

That's how Victor Brandl remembers the transformation of the land that took place with the coming of the pipeline. He has worked in the Canadian petroleum industry for fifty years. A wandering roughneck, he came to Fort St. John in the drilling boom that followed the completion of the pipeline and seized the opportunities he could see ahead. He created two oil field construction service companies: Northwest Oilfield Services and V. E. Brandl.

Before Frank McMahon's drill bits punctured the sediments beneath the obstinate crust of the Peace River country, that rolled off the east slopes of the Rocky Mountain ranges, it was a remote place studded with the tepee rings and campfire remnants of twenty centuries of prehistory. Following the original contact of the Europeans with the aboriginal inhabitants, it remained the dominion of the restless and the independent. Explorers mapped its river basins. Scientists of the Geological Survey of Canada sampled its soils and rocks, and recorded its flora and fauna, then scurried back east to more benign surroundings.

The twentieth century laid a light hand on this place. Trappers still bivouacked along its creeks, sportsmen roamed the hills and homesteaders broke its sod. Only the drone of bush planes shattered the silence; only a few frost-rutted roads crossed its valleys and probed its mountain passes. The character of the geography shaped the character of its people: self-reliant, adventuresome, resourceful, hardy and hard-living. Isolation rendered them wary and secretive, but they were also considerate and neighbourly. Survival is always a collective enterprise in these latitudes.

When the drilling rigs first arrived after the Second World War, followed by the hurly-burly of the pipeline spreads, the unyielding land branded the new frontiersmen with its harsh qualities rather than altering its own character. Narrow seismic lines and service roads threaded the landscape, wellheads appeared on the hillsides, 60-foot pipeline rights-of-way swept up the mountain passes; but these features were swallowed up in the vastness of the landscape. This slab of virgin wilderness remained a world away from the cocktails and contrivance, the dealing and duplicity of downtown Vancouver or Calgary. The pipeliners who laid the steel and erected the plants were a breed apart from their corporate counterparts who, from 1947 to 1955, had wrestled a binder full of permits from furrow-browed bureaucrats and inveigled $170 million in financing from jittery brokers, skeptical bankers and pinchpenny insurance investment managers.

Westcoast's formation, and the initial struggle to build its core pipeline, went to the very root of Canada's classic dilemma: how to obtain enough capital and know-how without becoming indentured to its agents in the process. For once, the formula worked. And it worked because behind the cohorts of lawyers and accountants who put the deal together, there came a legion of welders, cat skinners and sweat hogs who put the steel in the ground and then got it to do the job for which it was built. To accomplish this, they repeated the five-hundred-year North American cycle of vanquishing the land. But they did so with a lover's hand, for the land in its turn vanquished them.

THE MEN AND WOMEN who built the line belonged to the mountain passes and rivers, the swamps and wheat fields, the brooding forests and moody coast lands that they secured, breached and subdued for the original right-of-way. Many were pipeline nomads who wandered the continent after the war like migrant harvesters with their fleets of side boom tractors, trenchers, pipe wrappers, welding machines, ditchers and backfillers. They were rye whisky drinkers who thought that a martini was an Italian village. They lived a rough life and had a sense of humour that fit their style. Dewey Lund, a labour foreman in 1957, recalled what happened when a bear appeared on one work site:

> We were building a water line, in a ditch about 10 feet down. I had a bunch of guys in the ditch wrapping the pipe. Well, this goddam bear comes along and went down into the ditch, wanting to cross to the other side. But he couldn't get back up, so he started running down the ditch and, boy, you should have seen those guys' arses going up out of that hole ahead of him!

When the construction crews were done, some four hundred men and women settled in B.C.'s Interior with their families to run, with commendable efficiency and safety, the system of gas fields, processing plants and pipelines that became Westcoast's operations' heartbeat and financial lifeblood. These permanent corporate yeomen built homes and cottages, sat on school boards and town councils, joined hockey teams, married their local sweethearts, raised children and lived out their professional lives under twin shadows. The pipeline and its distant links to the fiscal furnaces of Vancouver and Seattle cast one shadow;

the mercurial contours of the Peace River country, stubbornly retaining its frontier flavour and climate, cast the other.

However brilliant, however visionary, however intuitive, the executive leadership of a great energy company stands or falls on the ability of its field personnel to perform the leadership's bidding. At Westcoast, the pipeline gang delivered the goods unfailingly.

FOR TWO FRENETIC SUMMERS, the pipeliners gouged a continuous six-foot trench through forest and swamp, across wheat fields and grazing lands, under creeks and streams. It had taken twenty years of planning, ten years of gas well drilling and six years of legal and financial manoeuvring to get the first shovel into the ground. In contrast, the speed with which the pipeliners got the big-inch line in place and completed the compressors to drive the underground train of gas and the upstream gathering and processing system to supply it amounted to a mere twenty months. "As it turns out so often, construction was not the time consuming part of the project," Frank McMahon remarked laconically in his 1956 annual report, putting seven years of regulatory manoeuvres into wry perspective.

Officially, work started as soon as the frost was out of the ground in the spring of 1956. Actually, construction kicked off, shrouded in secrecy, on October 21, 1955, at the international border near Huntingdon, B.C., against a backdrop of golden aspen and poplars and a grey sky as the chill autumn rain soaked through the men's windbreakers and thick mud caked onto their boots. When the first ditchers cut the first miles of trench, Westcoast still lacked a pair of key permits: the Canadian and American governments' export and import certificates to move gas across the border.

However, Westcoast had taken delivery of the first order of 30-inch pipe, which had been shipped from Britain to the Port of Vancouver that summer. Steel pipe was the only commodity harder to come by than government permits in those years of non-stop pipeline construction around the world. Westcoast placed mill orders months ahead of schedule, knowing that in the worst case, if they took delivery and the regulatory system let them down again, as it had in 1954, the pipe could be resold. However, with the uncertainties of steel supply during the Korean War, even with solid orders lined up, McMahon feared that when he had his approvals he might still lack the pipe. In San Francisco, legendary construction executive Steve Bechtel had a solution. Head of Bechtel Corporation, the engineering and construction giant that built pipelines, dams, airfields and power plants worldwide, Bechtel provided his Canadian subsidiary with an option on 96,000 tons of 30-inch pipe manufactured in the South Durham Mills of England for another Bechtel project that had fallen through. McMahon bought the pipe and had it shipped to Vancouver.

In the autumn of 1955 with tons of pipe off-loaded and the needed permits dangling just beyond reach, Westcoast took a gamble, completely in character after surviving years of regulatory intrigue. The pipe needed to be stored over the winter, in the expectation that the final approvals would be in place for a full construction season in 1956.

Frank McMahon's trusted lieutenant Pat Bowsher took the initiative. "The safest place for this steel is six feet in the ground," he said in one of pipelining's immortal moments—an equivalent to General MacArthur's "I shall return." Somehow, he persuaded the Royal Bank's oil and gas lenders in Calgary to discreetly advance Westcoast $19 million in spare change in a handshake deal, and he got Alberta construction magnate Fred Mannix to quietly assemble a small crew of trusted, tight-lipped men. Charlie Hetherington thought it would be discreet to construct this "pre-build" away from scrutiny, up near Fort St. John. But McMahon insisted that they do it where everyone could see it and realize that the pipeline was going ahead. And he buried the pipe—after welding and wrapping it to pipeline specs—along several miles of the planned right-of-way from the American border up into the Fraser Valley. The work was already completed, a month later, when the final permits came in from Ottawa and Washington. The company did not admit to this enormous risk until Westcoast published its first annual report. Had the manoeuvre been unmasked, angered regulators might have killed the long-awaited permits.

The Westcoast main line, under construction following Pat Bowsher's and Charlie Hetherington's end-run in the autumn of 1955, had a precursor: 18 miles long, 4 inches in diameter and built of pipe salvaged from the abandoned Canol oil line. Westcoast completed this line in October 1950, bringing a trickle of gas from Pouce Coupe to Dawson Creek. The line had some notoriety at the time for being the most northerly gas line in the world. It was most important, however, as Frank McMahon's toe in the door to an integrated gas production, processing and transportation system to bring large volumes of Peace country gas through nearly seven hundred miles of magnificent and rugged interior country to the Lower Mainland and the U.S. northwest states. He and his confreres at Westcoast had chafed during the five additional years that it took to get the American and Canadian regulatory approvals needed to finance and construct his dream.

5

Building the Line

This is the greatest event for British Columbia since completion of the Canadian Pacific Railway united the province with the rest of Canada.
W.A.C. BENNETT

In the chill spring of 1956, with the growl of trenchers and the zap of welding torches at four construction spreads across British Columbia, the waiting finally ended as the earth opened and the pipeline took form from the scattered lengths of steel. By freezeup, the line was 70 percent finished. The network and plants and compressors to gather field gas ran ahead of schedule. The crews building compressor stations and the plant coped with extreme frost and severe blizzards to continue work through the winter—the tool of choice was the steam hose. The executives in Vancouver assembled the final pool of capital needed for the push to completion in the fall of 1957.

Along the line, the work of getting the pipe into the ground, then preparing it for operation, was dirty and dangerous. The massive equipment could crush a man in a second; it could bog down in swamps; it could slide out of control on the sides of hills or riverbanks. To build the aerial spans that Westcoast invented for the major river crossings required the men to dance along the suspended pipe like ironworkers installing steel beams for a skyscraper.

Dewey Lund remembered putting one water line underneath a road: "We never had those fancy drills and whatnot they got nowadays. We got down on both sides of the highway and dug by hand, tunnelling until we got the hole twenty feet in from both sides. We sent one of the hand's dogs through the hole with a rope towing a cable. Then we sucked the pipe through with the cable and just cleaned the dirt out of the pipe once it got through."

Getting a completed line ready for use involves several steps that are as hazardous as building aerial spans. One is the hydrostatic test—running water at high pressure—on a completed section of the line, to ensure it has no leaks that might burst when the first natural gas comes through at high pressure. If the pipeline passes the test, it is then purged: liquid methanol removes traces of water, and the air is then removed and replaced with gas.

Before the hydrostatic test, the crew must physically clean the sides of the pipe and remove debris that may have been left behind. Often bits of welding rod and even hand tools are forgotten when the pipe is welded and lowered into the ground. Tubular scrapers are pushed through the line with 25 to 50 pounds of air pressure. As they rush through the pipe, they emit a squealing noise for which they have been nicknamed "pigs" and the operation "pigging." Pigs are the mascots of pipelining. One tradition that developed in the 1950s was to name pigs after scantily clad pinups painted on the sides, just as bombers had been christened during the war. Gate-like devices called scraper barrels or pig traps are installed to insert and remove the pigs.

Bernie Guichon remembers the dramatic day when a hydrostatic test went wrong and nearly killed four men. A pig trap had been improperly installed at the foot of a 200–300-foot slope down which the pipeline had been laid. Adjacent to the pig trap was a line valve, and a crew of four men waited to operate it manually during the hydrostatic test. They listened as the column of water, followed by several hundred gallons of methanol, came roaring down the hill. The water slammed into the mis-installed pig trap, which acted like a brake, so that the backlash of pressure burst the trap:

The four men were manning the valve to let this water through more slowly. When that wall of water hit, there was nothing they could do. It opened up the trap just like it was made of toilet paper. They could have drowned; that they weren't seriously hurt you can't explain. One man lost a tooth. There was an Italian truck driver with a tanker truck to catch the methanol. He was sitting in his truck and the wall of water carried him out into the lake at the bottom of the hill.

The methanol, of course, wasn't caught, and it flowed out into the lake, too. An old man named John Logerquist, who had an Esso service station there, thought his gasoline tanks were leaking because he could smell this.

The doctor who looked after these fellows was on call at the Williams Lake Hospital. He had just delivered my first-born so I knew him quite well and he knew what I did for the company. Well, he treated these men and then late that night he called me. "I don't know what to say to you," he said. "Those four pipeliners were drunk and could have killed themselves."

I thought about it and then I realized—this was the methanol. They had inhaled enough of the methanol to be drunk, but that wasn't from drinking on the job. Fortunately, the doctor talked to me first, otherwise, if it had got in the paper that these men were drunk, there would have been hell to pay.

The perils of the project weren't confined to the pipeline; the men building the system for gathering gas from the wells and the processing plant at Taylor were exposed to a jeopardy that was unfamiliar to Canadian operators: sour gas. Most of the raw gas contained hydrogen sulphide—H_2S—which is toxic in very small quantities. It has a distinctive rotten-egg smell, but it is so fast-acting that, encountered unexpectedly, especially in confined spaces, it can knock a man out before he realizes it's there. Now, sophisticated gas detectors around plants and pipeline installations can warn of the danger, but the standard precaution is to wear an air pack—a mask and tank of breathable air—when working in a hazardous area. Ron Murphy started with Westcoast as a jack-of-all-trades, and soon found his niche as a technical service specialist in and around compressor stations, gas liquids and product pipelines. He recalls:

> We had a lot to learn about sour gas. Westcoast was the first company to run a sour gas gathering system in Canada. I guess in that first season every man including myself was knocked down on sour gas. No one was familiar with how to work with it, or handle it or cope with it, so we went to the school of hard knocks.
>
> In one episode, I had a mask on, but we were new to masks. I bent down to pick up a 26-inch-long pig and it caused the mask to fall away from my face and I got a whiff behind the mask. I had gotten a whiff and was backing away from the danger zone—and the next thing I knew my partner was taking my air pack off. I'd passed out and he'd come to rescue me.
>
> . We learned that, in that situation, what you did was open an air valve and blow the gas out from behind your mask—but we gained that knowledge by getting knocked down and monkeying around and figuring out how not to have something happen again. We had a lot of Yankee supervisors who'd built plants, but they hadn't operated in a sour gas environment either, so it wasn't a very good situation as far as working conditions.
>
> Thankfully we had no fatalities, but everyone had a brush with sour gas; and we passed on to each other what we learned—and after that first season there were no gassings for nine or ten years. Then a flock of new people came on board, and we didn't pass our schooling down effectively enough, so there was another cycle of three or four people that got gassed and knocked out, until we got the teaching right.

Some brushes with danger were more comical than risky. Area foreman Bill Crutchely remembers a crew of local Native men was hired to scrub off pigs, pig traps and pig barrels along the line. It was summer, and one safety precaution for isolated crews on the pipeline was to have a man with a shotgun act as bear watch.

This particular crew was fifty or more miles from nowhere and these kids are out there painting and scraping, and Johnny Clark is the bear watch, but he falls asleep leaning up against a storage shed. He was sound asleep, and he felt something on the side of his face. It's a damn black bear, and he wakes up to feel this thing sniffing his right ear.

Well, he never fired a shot; he turned and saw this creature about six inches away—a great black bear's head looking him over. Old John just dropped the gun and ran like hell; and the bear, he ran like hell too, in the other direction.

Laughing in the face of danger became a way of life; Westcoast was either remarkably blessed or plain damn lucky in that few incidents had serious consequences. So the stories of the more spectacular accidents became company lore. Stories circulated and were embellished. An electrician fleeing an incipient explosion, when a gas leak was detected inside a building under construction, scaled an eight-foot Frost fence. In the retelling, the fence grew taller, and gained a topping of barbed wire, and he didn't climb it, he leapt it with one bound, and later was blown over it by the force of the explosion.

The outcome for Westcoast, however, was that in the tight-knit cadre of pipeliners and operators, men valued the lessons learned in dangerous moments. There were no fatalities during construction, and no serious accidents in terms of property loss, death or disabling injury for several years. This was the best job that most of the pipeliners had ever had. The company was barely out of its gestation, but it was legendary for the patience and vision of its revered patriarch. These elements contributed to a protective culture along the pipeline. The people of Westcoast became a family. The men became safety-conscious not just to protect each other, but because they savoured this new way of life. Westcoast had remarkably few serious injuries, and over time that fact transferred itself into an ethic of environmental protection in which all life—wild and human—was respected. In spite of the many stories of narrow escapes, Westcoast became one of the safest resource operations in British Columbia.

WHAT HAD WESTCOAST'S 2200 men and women wrought, one ton of dirt and one length of pipe at a time? They had constructed an energy subway delivering 400 million cubic feet of natural gas per day, two-and-one-half times the capacity of British Columbia's hydroelectric power complex in 1957. They had completed the world's first international natural gas grid, shipping 300 million cubic feet of gas per day into the Pacific Northwest states, an innovation of corporate and political diplomacy that initiated the continental energy market thirty years ahead of its time.

The Westcoast Transmission right-of-way ran like a steel spine through British Columbia's heartland, binding the remote northeast to the Lower Mainland, remaking the mercantile anatomy of the province. The route simmered up in the placid wheat fields of the Peace River, where a trick of climate created a pod of land amenable to agricultural production. It climbed the low elevations of the Rocky Mountains' eastern slopes, up through the gentle incline of Pine Pass. It travelled through dense forests and into the cattle country of the central Interior, where million-acre spreads of parkland and grass nourished great herds. Trekking south, through the central drylands, it swung west at the Coquihalla Pass into and across the Cascade Mountain Range and plunged down to the Fraser River Valley, following that artery of river and railway commerce past sand bars once panned for gold dust and out past the salt-kissed delta farms of the coast to the international border at Huntingdon, B.C.

To lay this industrial sinew across an expansive province during two construction seasons in 1956 and 1957, the pipeliners dug and blasted 7 million tons of rock and soil; they welded together and interred 230,000 tons of steel. They installed 41 railway and 66 highway crossings, negotiated 70 small streams and creeks and 7 major rivers, and strung out 5 breathtaking aerial suspension bridges to carry the line across the most difficult rivers. They installed 52,000 horsepower of compression to move the gas. Above ground, their monuments were four compressor stations at 160-mile intervals along the line, and the sprawling gas-processing plant on the northern end of the line at Taylor, B.C. Below the earth, the 30-inch main line artery threaded 650 miles through

British Columbia to the American border. At its upstream end, a web of 186 miles of small-inch laterals and 37 miles of 26-inch line gathered gas from 52 wells in eight fields on the British Columbia side of the provincial border and 30 wells in three Alberta fields.

After this system's completion in 1957, four hundred operators tended the 650 miles of main line and four giant compressor stations. Over the next decade, this number rose by six, and the web of gathering facilities spread out over 10,000 square miles. They also ran the $15 million Taylor gas-processing plant, which cleaned and dried the gas to pipeline specifications and produced enough acid gas to make 300 tons of sulphur a day, plus high-grade petroleum condensate liquids for, each day, 1560 barrels of aviation gas, 1700 barrels of motor and diesel fuels, and 500 barrels of propane.

Those are the numbers. In B.C.'s Interior, however, people measured the pipeline in an altered landscape and changed lives. The landmen who acquired the right-of-way had secured agreements with 1150 private landowners and fifteen Native bands, in addition to obtaining permits to cross nearly 500 miles of federal and provincial forests and grazing leases. Communities sprang up like flowers as the pipeline spilled its wealth across the landscape in a downpour of steady paycheques and annual property taxes, along with its unending hunger for the purchase of equipment, supplies and services.

The pipeliners settled down, built homes, raised families, found the local fishing spots, hunted every autumn, curled in winter and became the school board members, hockey coaches and mayors of the new towns that Frank McMahon's dream created.

FINALLY, IN THE FIRST week of October 1957, the week of his fifty-fifth birthday, Frank McMahon's big event—the pinnacle of his career—arrived. The last length of steel had been buried, the last electrical panel was wired up, the first wells were on stream, the processing plants were up and running, the valves were tested and certified. The party to launch the Westcoast Transmission pipeline had taken three months to plan and prepare. It ran non-stop for three days. By the time

previous page: From October 21, 1955, to October 7, 1957, an army of 2200 pipeliners welded and buried the 1000-kilometre Westcoast Transmission natural gas pipeline from British Columbia's Peace River country to its Lower Mainland and the American border. WESTCOAST ENERGY

above: When the pipeline first went into operation, the Taylor gas plant near Fort St. John processed production from fifty-two wells in eight fields in northeastern British Columbia and northwestern Alberta before it was shipped south. WESTCOAST ENERGY

left: Crews building the pipeline crossed the toughest terrain in North America. A pipe-wrapping crew works on a precarious perch in the Coquihalla Pass. WESTCOAST ENERGY

above: Premier W.A.C. Bennett (left) described the construction of the pipeline as the most important event in British Columbia since the completion of the Canadian Pacific Railway. He joined Frank McMahon (right) in Fort St. John on October 7, 1957, to turn the valves that started the gas flowing south. WESTCOAST ENERGY

facing page: The task of filling the pipeline with a reliable flow of natural gas belonged mainly to Pacific Petroleums, which Frank McMahon also controlled. Pacific Pete, as the company was called, sent its drilling rigs into the Peace River country to drill the wells to supply Westcoast's new markets. WESTCOAST ENERGY

facing page: A web of natural gas gathering systems snaked out of the pipeline's Fort St. John hub to wells spread across the Peace River country. Westcoast Transmission's insatiable thirst for gas would eventually push exploration and development into the Yukon and Northwest Territories. WESTCOAST ENERGY

above: The first foreign dignitary to tour the new pipeline was Britain's Princess Margaret in 1958, with Frank McMahon on a viewpoint overlooking the Peace River aerial pipeline crossing at Taylor. Westcoast was financially troubled, but McMahon basked in its accomplishments, confident that it would muddle through. WESTCOAST ENERGY

following page: Most of the pipeline was hidden from view, buried in rights-of-way that melted into the landscape. To conquer the Fraser River near Hope, however, Westcoast's engineers designed a dramatic high-level aerial crossing that called for courage and skill from the men who built the lofty link. WESTCOAST ENERGY

it was over, Westcoast had lavished $120,000 on the celebration, Pacific Petroleums had tossed in a few thousand more that no one took the time to add up, and McMahon had topped up the budget with $80,000 out of his own pocket. Western Canada had never seen such a splendid corporate affair, reflecting McMahon's acquired Hollywood and New York style as much as it did Premier Bennett's sense of the occasion's historical importance.

The festivities started on Sunday afternoon with a continental progress as a caravan of chartered airlines collected Frank McMahon's friends, the business and government officials he'd courted for eight years, and the planners and engineers who'd built the line. They came from New York, Ottawa, Toronto, San Francisco, Seattle, Vancouver and Victoria. They joined up in the best suites of Calgary's Palliser Hotel, where Pacific Petroleums hosted them at a dinner with plenty of toasts and a shower of gifts. In Fort St. John, a late autumn blizzard struck that afternoon, and as the party carried on far to the south, company crews cleared snow from the airfield, roads and sites for the next day's open house.

October 7 was brilliantly sunny in Fort St. John. When all the aircraft landed, the party numbered four hundred guests, who slogged good-naturedly through mud and slush to inspect the refinery and processing facilities and a gas well or two. In the afternoon, they gathered at the valves that marked the head of the pipeline and cheered as Premier W. A. C. Bennett, Frank McMahon and Westcoast division inspector Lin Bennett open the valves to start the line fill. Then they dined and drank for a second night as the gas flowed down the $170-million, 650-mile-long, 30-inch-diameter pipeline towards Vancouver. Through the night, the great artery pulsed with energy as the gas swept down the central Interior and turned towards the mountains and the coast.

The next day, in the Hotel Vancouver at a lavish luncheon, Frank McMahon, Premier Bennett, business associates, political cronies and the Vancouver aristocrats who'd made it happen marked the first delivery of natural gas to the Lower Mainland. A roll call of the most influential

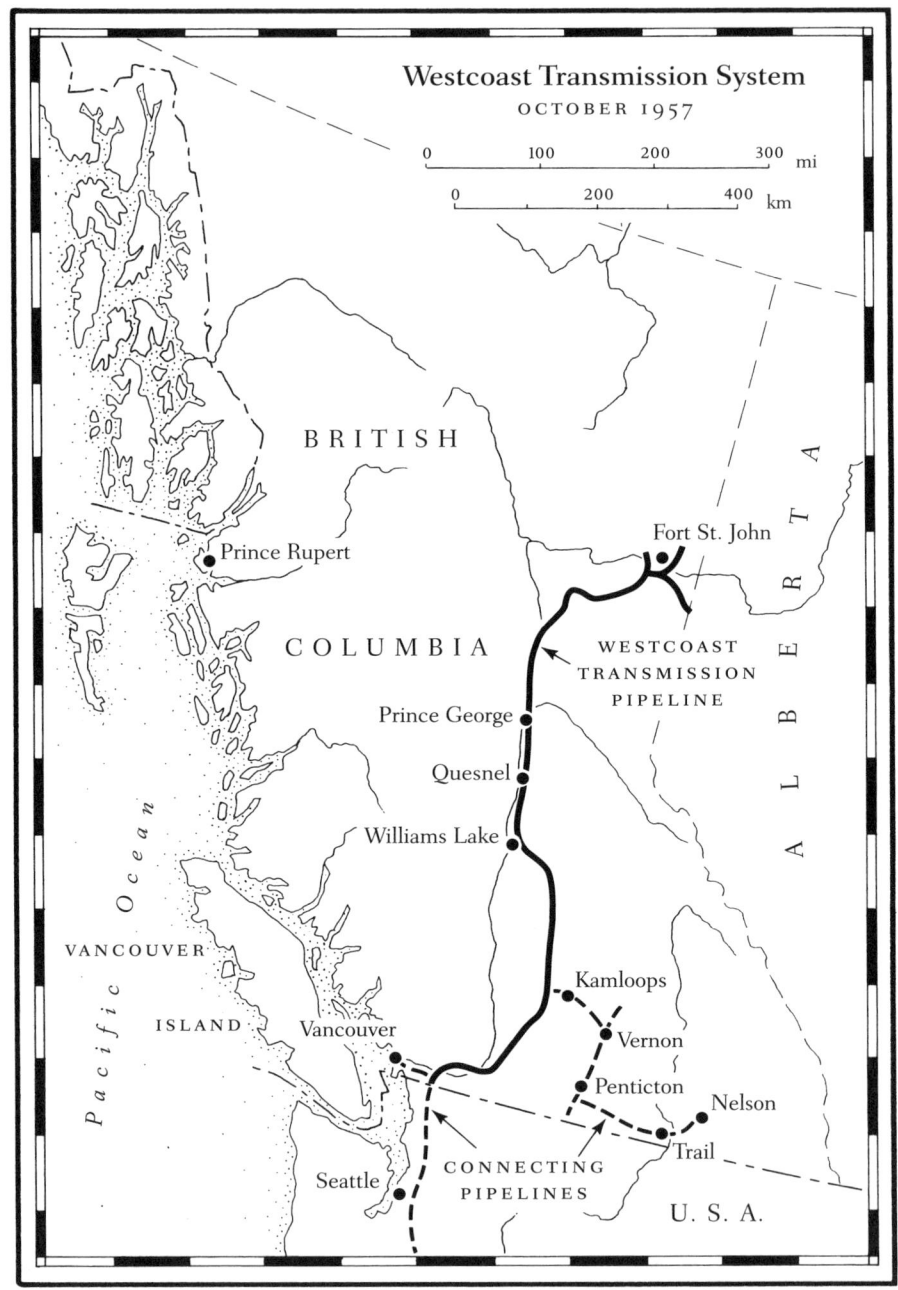

stood to praise McMahon. "This is the fulfillment of a twenty-year dream of the man who brought Peace River gas through most of British Columbia to Pacific Coast markets in Canada and the United States," said Dal Grauer, chairman of British Columbia Electric, which until that day had been the province's premier energy company.

Lest the jovial Westcoast executives succumb to the sin of hubris, the construction of the line faced its last glitch. Westcoast's enthusiastic PR man, former *Vancouver Sun* business editor Lloyd Turner, had rigged a microphone at the last valve on the Canadian side of the international border, to capture the sound of gas rushing into the United States. He had set up a public-address system in the Hotel Vancouver ballroom, intending to play the sound to the guests—giving the moment its piece of drama. At the last minute, the live feed from the pipeline site failed. Turner was obliged to fake it. He asked communications engineer to turn up the squelch on the two-way radio, generating a blur of static from the Hotel's sound system, passing it off to the admiring throng as the sound of gas running through the line—the moment in history for which they had all gathered.

That night, the celebration culminated with an enchanted gala in the formal ballroom of the Hotel Vancouver. Floral arrangements and artful lighting transformed the gilded hall into a regal setting. Four hundred well-fed Canadian political, business and financial leaders danced with stunningly dressed wives and mistresses, hosted by British Columbia's lieutenant-governor and long-time McMahon financial sponsor, Frank Ross. When the last weary celebrant drained the last brandy snifter and stubbed out the last cigar, the legend of the occasion had been indelibly stamped on the history of Canadian pipelines.

History always adds up to more than the sum of all the ascertainable facts. It is essential to record the first-hand experiences of the men and women who were there, making things happen not with any grand public gestures but by doing their grainy thing, performing the spade work that makes history possible.

The sinewy, rough-cut folk who built Westcoast Transmission succeeded in a highly complicated enterprise with bare hands, brute strength and skill at the controls of monstrous machinery. Their generation took a backwater wilderness in which resource extraction consisted of hauling fishing nets, springing traps and digging coal and converted it into part of an essential industrial complex.

The Westcoast people were the authors of their opportunity and their story echoes, fifty years on, with self-confidence and a sense of duty that drove them ever forward. The values and attitudes they espoused—and lived—continued to run like a steel thread through the tapestry of Westcoast Energy's character, and was the same corporate culture that gave the company its sense of an unlimited future.

6

The Force of Will

Westcoast people have something to brag about. They derive their unique pipelining skills from two circumstances: The early projects were generally under-financed, and the final costs had to be well below the same type of job elsewhere. And the pipeline traversed every type of terrain—permafrost, small glaciers, timbered mountains, high-water-table valleys, arid prairie, fertile flats, orchard country, granite cliffs of Pine Pass and the Coquihalla, raging rivers and the seismically active Fraser Valley. The topography, from the height of the Pacific Coast Mountains to sea level, was as tough as the geography.

ED PHILLIPS

When construction of the pipeline was completed in October 1957, Westcoast Transmission was eight years old and hadn't turned a dime of profit. Nor would it for ten more years. A return on a $250 million (and growing) investment had to be wrestled out of stubborn rock and bottomless swamp.

The sweat hogs out in the muskeg and bush, who replaced lawyers and financiers as the nuts and bolts of the enterprise, opened a technological frontier. No one in pipelining had achieved what the Westcoast people were doing, not anywhere in the world. Not even the 1944 Canol oil line from Norman Wells in the Northwest Territories to

Whitehorse in the Yukon had met everything Westcoast faced, and Canol was driven by the imperatives of war and abandoned in the first weeks of peace. Westcoast was first to take the singular challenge and risks of building high-pressure, large-diameter natural gas transmission lines and attempt them in a remote northern wilderness, contending daily with the combined hostility of climate, isolation, distance and terrain—not to mention blackflies and grizzly bears. An entire company was learning its business on the job.

Westcoast expanded and extended the system at a steady pace in its first years of operation, dancing to the tune that the flourishing gas market played and trying to grow fast enough to be economic. There wasn't enough capital. The shareholders never saw a dividend. The company was paying the mortgage with the grocery money.

The event that turned the chronic financial dilemma into a crisis was both unexpected and ironic because it involved a $16 million, 24-inch secondary gathering line on the far northern fringe of a major North American pipeline system that had soaked up nearly a billion dollars. It was built in late 1966 to bring raw gas in from the Yoyo field in the four-corner border country where British Columbia, Alberta, the Northwest Territories and the Yukon meet. In 1964–65, Westcoast built a $57 million, 30-inch pipeline extension linking new processing hub at Fort Nelson to the mainline at Fort St. John. It signed contracts with the gas producers. Now it had to fill the line and deliver the gas.

THE WAY BILL HARROLD remembers it, he was the only man "halfways sober" when foreman Jack Malcolm telephoned him as the Westcoast sweat hogs "partied up a storm" in Fort Nelson's Maverick Hotel. It was six in the morning, but these fellows could celebrate around the clock as easily as they worked twenty-four hours non-stop. They had just completed the Yoyo gathering line and the previous day pressured it up for its first test. The line failed, in the middle of the night in the middle of the party. And Jack Malcolm, "being the conscientious guy he was," had gone to the plant to check the pressure values, though anyone else might have given the test a few hours.

"'Bill,' he told me," recalls Harrold, "'we've got to go flying in the helicopter to find the leak. The pipeline doesn't have an ounce of gas in it.' That's when it started. The pipeline was faulty and it kept blowing up. We had fifty-six blow-ups or something like that before it was abandoned. It never did reach its correct pressure. I've seen areas with 400 feet of line blown out of the ground, split right open and twisted like a rope."

According to Ed Phillips, Westcoast Transmission's former chairman, president and CEO, the company had no choice but to work with the troublesome line.

> The reason it never got into service was that the company didn't have $1 million much less $16 million to start over. The men had to try repeatedly to replace the blown sections. The engineers wanted to call a halt and rebuild. Doug Owen pleaded. Frank McMahon said, "We don't have any money. Repair it, it'll work next time."

Bill Harrold, who started with Westcoast as a payroll and accounts-payable clerk during the Fort Nelson project's construction, gained the reputation of being the man who knew the details about the pipeline that everyone else had forgotten. He participated, every step of the way, in the two-year attempt to get the Yoyo line working. He recalls:

> We would test it with gas in the winter but with water in the summer, otherwise the line breaks would start forest fires. We had a pool, like a hockey pool, on what the pressure would be when it blew next time. It was supposed to go up to 1500 pounds per square inch, but every time we got it over 900 or 915 pounds, it would let go. Sometimes it would blow six or seven times in one area, and we'd have the muskeg dug up for miles until it was just one big swamp. Or we'd run into permafrost. Usually it was 14 feet deep but sometimes it came up to 3 or 4 feet. We couldn't trench it, so the helicopter would sling in a compressor and a jackhammer. We tried to dynamite it, but we hit the pipe every time.

We usually cut out pieces of the line to repair them, but once we had a guy named John who was small enough to get inside a 24-inch line, and he said, "Don't bother." We tied a line for safety around his leg, and he got a mechanic's creeper and took his welding torch and scuttled down inside the pipe, and repaired it from the inside. It was a no-no to be welding inside a gas line, but you did what you had to do.

We lived rough working on the Yoyo line; we had to bring things in by helicopter or on dual-track Foremost vehicles. We worked sixteen hours or around the clock and slept in a tent or just under a tarp. For showers we had a ten-gallon barrel with holes punched in it, then bunged. We'd fill it with water and let it warm up in the sun. Then we'd open the bungs and have a shower.

Finally, we abandoned the original pipeline and put in a new line right alongside it. Actually, you know, there's still pipe lying out there that was blown up. There's quite a bit of bush growing up around it now, but you could probably still spot it from the air, if you went in by helicopter.

The continuous cycle of repair, rupture and repair had cost the company time, money and credibility. Their lawyers sought a remedy. Ed Phillips remembers that:

When we stopped repairing the line, we sued the pipe maker. The judgement was there were non-metallic inclusions in the longitudinal spiral welds along the pipe. These impurities reacted to the H_2S that, of course, was to be cleaned out at Fort Nelson before the gas was sold. In a matter of hours, the sour gas would attack the impurities and create a pinhole that, when you put pressure on it, would explode. The lawsuit was settled out of court for about half the $13 million Westcoast claimed, and it wasn't sufficient to square the account.

The men and the company were glad to be engaged in more productive work. The Fort Nelson processing complex and the big-inch connection to Fort St. John had been a success in spite of missing the gas from Yoyo. Other fields made up the supply gap. The financial dilemma the company faced with the failure of the line went far beyond finding $16 million or so to replace it. Westcoast had signed "take-or-pay" contracts with producers for the gas in the Yoyo field that was supposed to be moving on the line. So it was writing cheques for gas that was staying in the ground. The last thing cash-strapped Westcoast needed was a frozen inventory.

El Paso came to the rescue. If Westcoast Transmission was dependent on its U.S. customer, now purchasing 70 percent of the gas delivered on the main line, El Paso was also dependent on the Canadians, who provided it an irreplaceable volume of gas. The last thing El Paso needed was the financial ruin of its northern supplier. So, it stepped up to finance the replacement of the failed gathering line, with a common share financing and an understanding its export supply contracts would be protected.

THE EL PASO FINANCING solved one problem but created another. As the gas from Yoyo flooded into the pipeline, the North American economy sagged. Gas sales to Westcoast's principal customers, BC Hydro and El Paso, did not grow as fast as anticipated.

Westcoast wasn't able to sell all the gas it had purchased under the "take-or-pay" terms of its contracts. It needed a bigger market, and moved now to bring the long-delayed gas line from Prince George to Prince Rupert to life. The solution was characteristic "Frank McMahon." He simply told his engineers to design a pipeline to carry natural gas from the main line at Prince George. He said he would form his own distribution company (a retail operation that was new to him) to sell the gas. There were few residential customers to be found along the way. Possibly small feeder lines could reach a few isolated mines. Mainly, he had his eye on the forestry industry at the end of the line. With a good portion of debt, he raised some equity capital for a new company called Pacific Northern Gas, in which Westcoast would hold all the voting stock.

But, when the tenders for the 10-inch pipeline were opened, the lowest cost, $36 million, was considerably beyond the most optimistic revenue estimate from the gas to be sold. Undaunted, McMahon simply told his engineers to redesign the pipeline at a cost of $28 million to fit the revenue stream.

The engineers knew that conventional construction plans could not reduce the gap. But they knew the boss would not accept a negative report. They presented a plan to reduce the diameter to 8 inches on the western leg of the line from Terrace to Prince Rupert, and decided to string the line uncovered along mountains, muskeg and rock across the most difficult terrain, saving the immense cost of ditching to bury the line.

The engineers explained to the chairman that this was the only plan that would reduce the costs enough to make the project viable, fully expecting and even hoping he would reject the fanciful scheme out of hand. After the long fight over the abandonment of the Yoyo line, one might have expected these men to be more wary of their employer's out-of-the-box thinking. He applauded their resourcefulness and immediately authorized construction of the dicey proposal. In the end, the company spent $32 million on the project.

Keith Irwin, who worked as the land agent and construction inspector on the project then stayed to operate the line for fourteen years, says, "the world's worst pipeline country runs from Terrace to Prince Rupert. The pipeline follows a river up to the timberline, then passes through a bored tunnel 700 feet and goes down the other side. In places, it hung by cables on the rock face of cliffs. Some portions were secured to large rocks or trees for stability on slopes."

Bernie Guichon recalls the haste with which the project was completed:

> We built that whole system in a year, 500-and-some-odd miles plus the lateral lines. It was buried on the south side of the Skeena River. On the north side it was just rougher. That's the real Coast Mountain area. It was so swampy and yet so rocky that you couldn't work with

it. Strangely, when the right-of-way got cleared, it seemed to dry out and it got so that you could work with it and keep the line running, especially in winter.

The crews who built the line literally chained it to trees, anchored it with blocks to some slopes and let it float in some muskeg lands. They got away with it because the right-of-way was extremely remote. The few trappers who might have to climb over or under it didn't care one way or the other.

For years, when the B.C. government inspectors conducted overflights of the system, the practice was to distract the inspectors when passing the exposed portion of the line. This section of the tour was left to the last, after the tedium of flying over a couple of hundred miles of buried line. Someone always broke open the wet bar and poured drinks during the time that the aircraft traversed the naked pipe.

The situation gave the operating crews problems for several years. They inspected the line unceasingly. Each year they found money to bury small sections of the line, slowly getting it all below ground. And they had incredible luck. They turned an impossible situation into a reliable piece of pipeline. In spite of line damage and the need for difficult repairs, no leak ever led to serious personal injury, permanent environmental or property damage, or financial injury.

Maintaining the Pacific Northern Gas pipeline over the years has been an awesome task. Ron Murphy started with Westcoast in 1958 and transferred to Pacific Northern Gas to supervise technical services:

> There is one section they call the bowling alley, which is on a sheer mountain face where ice boulders and rocks come down, and they have punctured the line two or three times. You can't get equipment into the bowling alley, so repairs are done by the arm strong method. You go and bull things through by pure strength and determination. Sometimes floods wiped out long sections of the line; the record was an 11-mile segment.

However, says Ron Murphy, the conditions are so rugged that maintaining the line defies normal practices. In the spring of 1999, the Skeena River crossing scoured out in the most recent of a series of failures. The crossing is difficult to maintain due to the erosion power of the river. The water eats into the riverbed much more deeply than a conventional pipeline trench. The maintenance crews have plenty of experience laying in temporary lines and repairing the damage; directional drilling techniques have advanced in recent years and will allow the company to get a line deep enough below the river to end the problem.

Even with the bare-bones design, the initial revenue was inadequate to support it, and when the British Columbia government refused to increase gas rates, Westcoast was forced to subsidize the operation for some years. The financial squeeze required some masterful management by the man put in charge of the problem child, Ron Rutherford. Eventually the line's volumes grew to the point where the price charged for the gas covered the costs of construction and operation, and it stopped leaking red ink. With positive cash flow ensured, the company was able to secure the integrity of the line by putting the above-ground pipe into proper ditching. In the long run, the Pacific Northern Gas pipeline wasn't much of a financial success. Operating it was part of the price Westcoast paid for the rest of its British Columbia franchise.

7

The Leadership Crisis

History makes men.
FERNAND BRUDEL

In 1957, Westcoast's financial critics and pipeline competitors launched an unexpected attack on the upstart pipeline through the Borden Royal Commission on Energy. The fledgling government of prairie bombast John Diefenbaker created the commission. His suspicion of wealthy men and Americans was closer to popular Canadian attitudes of 1812 than 1957, and he was spoiling for a shot at big pipelines with U.S. connections. Westcoast's critics cared not a whit for Diefenbaker's politics, and may have detested them. The Prime Minister, however, was a convenient tool.

The Borden Commission, which began its hearings two weeks after the 1957 launch of Westcoast's transmission system, targeted Trans-Canada PipeLines. Approving and assisting the construction of Trans-Canada had brought down the Liberal government of Louis St. Laurent. Westcoast Transmission wasn't cited in the terms of reference; however, the upstart pipeline had its enemies and they intended to embarrass Frank McMahon. The commission, chaired by Toronto lawyer, financier and industrialist Henry Borden, set the stage for the country's first and most-effective-ever National Energy Policy and the creation of the National Energy Board to regulate pipelines and energy exports. On the way, however, it subjected Westcoast to a gruelling examination.

The commission challenged the price of gas paid by Westcoast's Canadian customers, compared to export prices. And commission counsel also questioned the 625,000 founders' shares issued for a nickel apiece to the first backers of the company, which were worth more than $5 when the pipeline was completed.

Frank McMahon and his backers had made about $3.5 million on paper with the founding stock, an embarrassing sum of money for a company that still hadn't turned a profit. The commission found that the stock transaction had been fully and fairly disclosed, and was a legal and acceptable business practice. But—said the commission—it hoped that others developing Canadian natural gas resources would not seek to exploit the opportunity for profit this way. It suggested that the National Energy Board should not approve oil and gas pipelines or exports if promoters took these kinds of profits on their equity. The board disregarded the advice and stuck to regulating energy, not the stock market. Westcoast took the public relations black eye without apology or comment, and the controversy faded.By contrast, gas price was a popular issue. On paper the U.S. export price was 22 cents per thousand cubic feet, compared to 20 cents per thousand charged to B.C. distribution companies. However, Westcoast assessed additional contract charges that lifted the Canadian price to 32 cents per thousand. The prices were determined by fixing a common price at the city gates of Portland, Seattle and Vancouver, then deducting the cost of transportation from the American border. Pacific Northwest buyers drove a hard bargain back in 1954, but at the time the contracts were drawn, the Americans had the upper hand. Without their gas purchases the pipeline would not have been built. By the time the Royal Commission issued its findings, the gas price issue had evaporated; British Columbians were busy buying gas stoves and furnaces with the rising salaries they were earning in the province's resource development boom.

The Borden Commission and its newspaper coverage inflicted little damage on Westcoast's image and did not affect its bank accounts at all. But it was a warning shot, a precursor to the leadership crisis that was coming. The message was that the company's founder and its leadership were not universally loved.

The Royal Commission was just a ripple on the water. Westcoast Transmission faced two serious crises from the moment the gas flowed—a shortage of cash and of gas supply. The combined impact put the company on the verge of financial collapse, a chronic condition it was to maintain for the next fourteen years. On the day the valves opened, the project had already missed key financial targets and was staggering under the weight of its financial and operational obligations. Probably as few as five or six men knew the underlying financial frailty of the pipeline or understood the precarious passage ahead. The problem was basic: the Founding Father was a great creator but a terrible administrator. The company needed a leader who knew how to make the pipeline make money.

Westcoast Transmission was not just a pipeline. Its ownership and management were interlocked with those of Pacific Petroleums and together the two companies formed Canada's first fully integrated natural gas operation, exploring for, producing, processing, marketing, transporting and exporting its product. The chain from the wellhead to the distribution points at the city gates of the Lower Mainland and Pacific Northwest was unbroken. With common control and leadership, Westcoast and Pacific were joined at the hip, to succeed or fail together.

Westcoast had no partner in the pipeline itself. Pacific Petroleums held gas supply contracts for 75 percent of the pipeline's volumes during its first three years, and it had to finance and drill the wells that would meet the obligation.

Westcoast had parcelled out some of the natural processing facilities at Taylor Flats, outside Fort St. John. A consortium of producers owned the scrubber that extracted petroleum liquids. A specialist service company, Jefferson Lake Sulphur Co., recovered and marketed sulphur from the sour gas. Pacific Petroleums held on to the refinery to produce aviation fuel and other high-quality products from the gas condensate. With the best pieces of the project under their ownership or control, Westcoast and Pacific Petroleums could pay—just barely—the huge bills for capital construction, but they could not make a profit.

Westcoast had underestimated its capital costs and was therefore chronically undercapitalized through the final year of construction. Pacific Petroleums was also seriously short of cash: when it started drilling up the natural gas supply required for the pipeline, no one accurately knew the amount of money it would take to complete the wells.

The pipeline could not make money without significant expansion because of the damage done to its economics by cost overruns. Fortunately, the opportunity to double the size of daily shipments was foreseeable; unfortunately, to seize this opportunity demanded even more capital investment. To become profitable, Westcoast and Pacific needed a partner.

There was one at hand; Phillips Petroleum, an Oklahoma-based American independent producer, had launched postwar exploration in British Columbia alongside Pacific Petroleums. In 1947, McMahon picked up petroleum leases 1, 2 and 3. Phillips acquired leases 4, 5 and 6. The two operators drilled almost side by side. When Westcoast issued requests for gas supply, Phillips signed on as the second-largest contract-holder next to Pacific Petroleums. Phillips participated as a fifty-fifty silent partner in the Taylor processing plant and refinery.

The common stake in the refinery brought the two companies into an indivisible relationship within the first eighteen months of the pipeline's operations. In the rush to launch the project, Westcoast Transmission had pushed through design of the refinery on the basis of gas samples from a handful of wells in the Fort St. John area. As the search for gas supply pushed north, supply was discovered and produced from wells with much different gas and fluids chemistry. The new feed stock fit poorly into the refinery's processing parameters. The plant would have to be modified if it was to produce high-octane automobile fuel and aviation gas, needed to make the processing complex profitable. In March 1959, Phillips put up the $5 million required and took over gas processing, refining and service-station retailing from Pacific Petroleums.

The intractable threat to Westcoast's financial viability through the first half of the 1960s was its continuous thirst for capital. In 1961 and 1962, it faced another expansion issue: the gas fields that Pacific Petroleums and Phillips had been finding extended farther and farther north. The company needed both a second gathering and processing centre at Fort Nelson, which would cost $19 million to build, and a 245-mile, $52 million extension of its big-inch transmission system. But the Yoyo financial crisis—including lost income as well as huge unanticipated costs—had pushed Westcoast to the wall in terms of its ability to raise capital.

To solve the problem, McMahon incorporated a separate company called Gas Trunk Line of British Columbia with the financial backing of its shareholders, Pacific Petroleums, El Paso Natural Gas and an El Paso subsidiary, Western Natural Gas. Westcoast's ubiquitous champion, the Royal Bank of Canada, contributed the initial $6 million in a major debt financing that enabled Gas Trunk to leverage the partners' equity. Nevertheless, Westcoast needed some good luck or a windfall to get it out of serious trouble.

The first break came, unexpectedly, from an initiative to expand Westcoast's influence in Alberta natural gas development. Frank McMahon had always kept a wary watch on Alberta. In the original plans for delivering gas from the Peace River country to British Columbia's Lower Mainland, he'd briefly considered an Alberta route. It would have run from Fort St. John east, then south along the eastern flank of the Rocky Mountains to the Crowsnest Pass in southeastern B.C., turning west to run across the bottom of British Columbia to the coast. That proposal proved to be shaky as a business proposition and impossible politically. However, Westcoast paid attention to Alberta and got a competitive jolt when San Francisco–based Pacific Gas and Electric, one of America's largest and most powerful utilities, developed a project to deliver Alberta gas to California through a pipeline across British Columbia and the Pacific Northwest. This was a direct challenge to El Paso Natural Gas, and indirectly to Westcoast.

Westcoast jumped in with a proposal to develop a newly discovered southwest Alberta sour gas play called Savanna Creek by building a $10 million processing plant and a $25 million, 107-mile pipeline Crowsnest Pass connection to the Westcoast Transmission system. McMahon incorporated Saratoga Processing Co. to execute the plan; although Pacific Gas and Electric built its project, the Alberta and Southern Natural Gas pipeline, Saratoga turned into a financial windfall. The cash profits that flowed into Westcoast's coffers almost from the moment the Savanna Creek facilities went into service in 1960 are to this day credited with playing a key role in keeping Westcoast Transmission financially intact through some of its toughest years.

In 1960, Frank and George McMahon, with an eye to retirement and to settling their estates, sold their controlling interest in Pacific Petroleums to Phillips Petroleum. Phillips began to capitalize Pacific's drilling operations and maintained a 45 percent interest as a result of a series of financings. Phillips also began a series of transactions, over the next thirty months, that allowed it to creep into unfettered control of both Westcoast and Pacific Petroleums. It crossed the Rubicon to complete control in 1963 with the acquisition from Westcoast's principal American gas buyer, El Paso (formerly Pacific Northwest), of its B.C. natural gas production, reserves and field facilities, plus the Westcoast Transmission and Pacific Petroleums shares owned by El Paso.

Frank and George McMahon retired from the executive management of Pacific Petroleums. Although Frank continued as its influential chairman until 1969, in his declining years at Westcoast he was more of an icon than a leader.

As Phillips Petroleum consolidated its control of Westcoast Transmission, Frank McMahon fought his last political battle on behalf of the company. In some respects, it was the toughest of the many he had waged because it pitted him against one of the company's staunchest allies, British Columbia premier W. A. C. Bennett.

Westcoast's gas transmission may have been unprofitable and its treasury strapped for cash, but there was another pipeline to build—at least in Premier Bennett's mind. The intensive drilling effort to find natural gas to fill the pipeline had produced an unexpected bonanza. Wildcatters discovered sweet light crude oil at Boundary Lake on the Alberta–B.C. border, in 1959. It was the first of a series of major fields discovered within a hundred-mile radius of Fort St. John. It brought a rush of new operators to the area. And it caught the attention of Bennett, who learned that unless there was a British Columbia alternative, the province's oil would flow to Alberta refineries. The premier found this intolerable, and he called McMahon to his office in Victoria for meeting after meeting, twisting the businessman hard to build an oil line along the Westcoast right-of-way to Vancouver's four refineries.

For months McMahon demurred. To be economic, an oil line required at least 25,000 barrels of production per day; the new fields were producing only about 8000 barrels. Bennett accused the producers of deliberately avoiding oil deposits in their drilling programs, and he warned McMahon that Westcoast owed British Columbia too much, in past favours, to renege on the oil line.

McMahon and Charles Hetherington scrambled to come up with a plan to satisify the premier; fortunately, Peace River country oil production was steadily rising towards the 25,000-barrel threshold. In 1960, McMahon tendered a plan to the premier for a 12-inch oil line along Westcoast's right-of-way from Fort St. John to Kamloops, where it would tie into the Trans Mountain Pipe Line Company's main line, which linked Edmonton to Vancouver. He also did the premier one last favour, campaigning for him in the election of 1960 against a strong challenge from the New Democratic Party.

Fresh from his electoral victory, the premier ensured that Westcoast's project cleared the province's regulatory system and avoided federal review—indirectly quashing competitive pipeline proposals, which depended on permits from the National Energy Board and Alberta. McMahon went one last time to El Paso and Pacific Petroleums to tap each of them for 12 percent of the equity of the Western Pacific Products and Crude Oil Pipeline Company, which was to build the line. He and his leadership team, including his brother George, Charles Hetherington, Pat Bowsher, Doug Owen and D. P. McDonald put up 18 percent. McMahon raised 10 percent in a public offering and Westcoast Transmission financed the balance of the project. Westpac built the oil pipeline in 1961; it was designed for a capacity of 75,000 barrels per day and opened for business delivering 27,000 barrels.

During the next seven years, exploration companies tapped into new oil field after new oil field. Although the discoveries were elephants, they exceeded expectations, and the line filled steadily until it was carrying close to 70,000 barrels per day by the end of the 1960s. During these lush years, Westcoast Transmission consolidated Westpac with a subsidiary called Peace River Natural Gas, which had been its oil and gas exploration and production arm, into privately owned Westcoast Production. In 1968, Westcoast Transmission president Doug Owen persuaded his chairman, Frank McMahon, to take the company public as Westcoast Petroleum Ltd.

Meanwhile, through most of the 1960s, Westcoast Transmission was in a dire financial straitjacket, and after Phillips assumed control in 1963, it was a grimly disappointed shareholder. The pipeline operation had consumed more than $1 billion in capital, but Phillips had never received a dividend. The underlying problem was a crisis in leadership. When Phillips gained control, it moved to correct the problem, parachuting in one of its senior vice presidents, R. B. Stewart, as the new president of Westcoast.

Bob Stewart was a cordial, gracious, competent man. He quickly became the most popular Westcoast executive-around-town. He gained respect among Westcoast's customers and the gas producers who supplied the line. The U.S.-based Pacific Coast Gas Association made him its first Canadian director. Stewart's Bartlesville, Oklahoma–authored mandate was to rein in an entrepreneurial runaway that had all the makings of a spectacular success but was on the brink of failure. Westcoast, as seen from Bartlesville, was an unruly teenager; bright, decisive men ran the company, but they operated as a alliance of mavericks, each responsible for an area of business, none working closely together. Stewart was to put the company through finishing school.

Stewart, however, got caught between the directives from Oklahoma and the proprietary instincts of Frank McMahon, the icon in the corner office who did not want to see his era end just yet. McMahon waited for his moment, then used his authority to demote the president.

In 1965, McMahon manoeuvred to replaced Stewart with his own man, Douglas Owen. Owen had cut his business teeth as a young executive assistant in C. D. Howe's office when Howe was Minister of Trade in the postwar St. Laurent government. After Howe's career ended, Owen moved to Calgary and parlayed his petroleum industry connections into a job at Pacific Petroleums. McMahon recruited Owen to Westcoast as the company's treasurer and assistant secretary; he had been promoted to executive vice-president during Bob Stewart's tenure. Although as its new president Owen oversaw the company's first Pacific Petroleums–mandated cost-cutting initiative, slashing the payroll by forty-two men, 20 percent of its complement, the company remained on the edge of financial collapse.

Westcoast Transmission needed a fix. When Frank McMahon turned sixty-five, he was eased out of the chairman's office. Doug Owen had proved an able president, distinguishing himself by creating Westcoast Petroleum and arranging the first mortgage financing in pipeline history to raise $40 million for the Gas Trunk extension to Fort Nelson, but he was seen by Phillips as too much Frank McMahon's creature. The hard men from Bartlesville sent in Kelly Gibson, chairman of Pacific Petroleums, as CEO. Gibson soon became known in the pipeline's Vancouver hallways—never to his face—as The Enforcer.

Gibson was a wildcatter, an oilman's oilman who came up right from the bottom. He had worked on the rigs in Oklahoma and other places in the U.S. and Alberta, moving from the field to the office because of his initiative, aggressiveness and work ethic, and his understanding of and appetite for the risks of the business. The oil patch hands at Stettler, Alberta, Gibson's last field posting before his promotion, took a less charitable view of the man, nicknaming him The Scorpion. They treated him with deference, however, because he enforced his authority with his fists, and had reputedly never lost a brawl.

He had no corporate experience and this showed in his rough management style. The positive thing about his tenure as CEO of Westcoast was that he could make decisions because he had the full backing of

Phillips Petroleum. Phillips sent him in to Westcoast to shake the place up, clean it out and make it profitable. Gibson reoriented the company; it stopped admiring the daring of investments and started measuring success by their profitability. Westcoast Transmission's shareholders, bankers and employees soon got the message that Westcoast was on a mission to become profitable, whatever the price.

An absentee guardian, Gibson was stern, disapproving, aloof, quick to anger and slow to praise. He never lived in Vancouver, instead flying into the city on his corporate jet in a sovereign style that generated its own resentments. The Vancouver executive regarded him, contemptuously, as a part-timer. However, Gibson believed his senior men should be at their desks six days a week. Only his strict Protestant belief in the Sabbath immunized him from mandating a seven-day workweek. Like many executives of his generation who came up from the oil fields where the job was never done, he treated vacations with suspicion. To ensure that his senior men were at their desks, Gibson called in every Saturday morning. At first they waited for the calls in their golf clothes or riding boots, depending on their preferred recreation and went out the door as soon as the accountability sessions ended. Later, they forwarded their calls to the twenty-four-hour operations centre, where the managers would cover for them, passing messages on to clubhouses and cottages so that the calls could be returned promptly.

The time Gibson spent in Vancouver constituted a continuous accountability session. He demanded results and imposed consequences for failure. He motivated by intimidation and humiliation. He boasted that he never fired a man. But, by shunning those who didn't measure up—demeaning them, freezing them out of decisions and sending them no work—he conveyed the message that he wanted them gone. When a pink slip had to be handed out, or a severance negotiated, that job was assigned to one of his vice-presidents. The thirteen executives deposed during Gibson's shake-up collected their pride and their severance and departed.

One man who didn't wait to clash with Gibson was Charles Hetherington. The swashbuckling engineer had worked briefly for Gibson at

Pacific Petroleums and had no intention of repeating the experience. He telephoned Ed Phillips from Calgary to quit, and didn't bother to return to Vancouver to clean out his desk. Hetherington went on to run PanArctic Oil with his legendary flair and some success. He was the only man from the founding team that put together Westcoast Transmission and Pacific Petroleums' near-monopoly of gas development in northeast British Columbia who could have succeeded Frank McMahon. His departure created a gap in the succession plan that Kelly Gibson was then developing—it removed the possibility of a future president and CEO who'd been present at the creation of the Westcoast Transmission system.

As Westcoast slowly turned course and steered towards the profitability that Phillips Petroleum wanted, it did so at the expense of corporate stability and executive morale. However, per share earnings multiplied ten times and dividends increased eight-fold during Gibson's tenure. Phillips Petroleum regarded the injured sensibilities of those who found Kelly Gibson difficult as the unfortunate consequence of their being spoiled by the indulgent, undisciplined McMahon.

Gibson clamped down hard on costs and stripped budgets until the massive cost reductions rolled out into the field, triggering layoffs. The red ink on the bottom line was stopped and replaced with ever-increasing gallons of black. Earnings and dividends soared. Financial analysts sat up and took notice. The relationship with the British Columbia government remained a solid partnership, even when the socialist New Democratic Party, led by Premier David Barrett, seized power.

Gibson expanded and extended operations; it is the burden of every successful utility and pipeline to maintain a constant capital construction program. But he chose his enterprises with calculation, not instinct. Under his regime, the company made its first successful initiative to reach outside British Columbia by participating with the Alberta Gas Trunk Line (later NOVA Corporation and now part of TransCanada Pipelines) in the Foothills pipeline project, gaining authorization to build a $26 billion natural gas pipeline from Alaska's Prudhoe Bay and the Canadian Mackenzie Delta.

Gibson created the first succession plan for the company's leadership worthy of the name. He built it around the small cadre of executive survivors, including Ed Phillips, who succeeded him. He recruited John Anderson, an Imperial Oil–trained petroleum lawyer who'd done outstanding work for Pacific Petroleums. (Gibson put Anderson out of mind as his immediate successor after the two men tangled. Anderson refused to put up with Gibson's bullying, and bridled over the contradiction between the clamp on executive expense accounts and the lavishness of Gibson's own lifestyle paid out of Westcoast's purse.)

There is an honour roll of men who kept Westcoast intact and moving forward through the years of crisis in the 1960s and the heavy-handed Gibson presidency. They were the leaders along the line, men who kept their distance from Vancouver's head-office skirmishes, kept focussed on the job and inspired the people doing the day-to-day grunt work that is pipelining in the rough. They included Vice-President of Operations Ed Johnson and supervisors such as Chuck Starr, Cec Pickell, Dick Littledale, Bob Merrill and Red Shannon. "These men were the near-heroes of Westcoast's defining years," says Ed Phillips, who in his own presidency came to fully appreciate that a pipeline isn't really run by the executive decision makers but by the hands on the valves.

Gibson's brusque management style during his eight-year guardianship was at odds with his face outside the company, where he was known as a gregarious golf partner, active service-club volunteer and devout United Church member who organized prayer breakfasts and United Appeal campaigns.

His reputation as a corporate director of such sterling businesses as the Royal Bank contrasted sharply with his treatment of the people under him at Westcoast Transmission. Northern explorer and promoter Cam Sproule wanted him to lead the PanArctic Oil consortium and considered Jack Gallagher, the first PanArctic manager, and later Charles Hetherington, its first president, to be distant second choices. Alberta Gas Trunk Line's Bob Blair ensured that, after he left Westcoast, Gibson became chairman of Foothills Pipe Lines Alaska gas initiative. And Gibson was to play a key role, as a veteran Pacific Petroleums director, in the dealings with Petro-Canada that led to the takeover of Pacific Petroleums and Westcoast by the Crown corporation in 1982.

Gibson did not like Vancouver, and the details of pipelining were unfamiliar to a man who preferred to find and produce oil and gas. He had a producer's ingrained distrust of pipelines.

Phillips Petroleum gave him a mandate to make Westcoast Transmission profitable, and that did not include instructions to cushion the blow on the people whose lives and jobs would change. Gibson carried out his orders to the letter. In the end, he did what he was sent to do and made Westcoast profitable, but the company needed more than that to achieve its potential. In 1972, when Gibson was kicked upstairs to become chairman, the board, the senior executives and the field operators unanimously wanted a more reasonable man, a conciliator and consolidator, to take his place.

8

The Great Conciliator

If I'd have done the research on Westcoast Transmission that I was accustomed to doing on a company, I'd have never gone to see Frank McMahon. But I came to work for a hero and I saw beyond its problems what Westcoast could become.

ED PHILLIPS

At first, the white collars at Westcoast Transmission's head office in Vancouver thought the unassuming man sitting in the barren office down the hall from president Doug Owen's suite and ploughing through files was the grim reaper from the Royal Bank. No one introduced Ed Phillips around the office. He simply arrived after lunch one day with Frank McMahon, and spent the rest of the afternoon closeted with Owen and the company's legal overseer, D. P. McDonald. When he later ventured out of his office and began to walk around, introducing himself, the beleaguered colony of harried managers struggling to save the company from its flight into financial ruin realized that this lean, watchful fellow was here to stay.

His profession and his new job at Westcoast were both vague. He denied being a lawyer, accountant or engineer, describing himself as a generalist. He said the chairman had hired him to help build new pipelines out of the north. Because the oil and gas to fill such lines was

American, from new fields in Alaska, and the economics—even for a healthy company—almost insurmountable, this assignment was baffling. Was this mysterious refugee from Toronto's branch-plant industrial economy here to play Sancho Panza to the Don Quixote in the chairman's office who could not kick the habit of dreaming impossible dreams? What could this stranger possibly have to do with the company's only important project: survival?

Edwin Phillips signed on at Westcoast Transmission, in 1968, as a fifty-year-old vice-president embarking on a third career. No one, least of all the man himself, assumed he would, in time, bring the company back from the brink of failure and shape it into a dream realized. There were other people ahead of him as candidates to lead the company. He had no pipeline or natural gas experience, and he was as unlike Frank McMahon as an executive could be.

In contrast, his career to that point was a perfect foil to McMahon's. He had advanced by standing in the shadow of great executives and learning from them the knowledge he lacked because he hadn't received a formal business or professional education. Ed Phillips deliberately explored in the mainstream. He fit the "man in the grey flannel suit" stereotype in every detail.

When he walked into the bar at Vancouver's Bayshore Inn, on the strength of a banker's introduction, to meet Frank McMahon for the first time, Ed Phillips was intending to pass a pleasant lunch with someone he had long wanted to meet. At the age of nineteen, he'd thought about living in Vancouver but ended up in Toronto. He rose through a series of postwar executive positions at Loblaw Groceterias, Dominion Sugar, Consumers' Gas and Trane Canada, the heating and air-conditioning manufacturer where he'd been promoted steadily over sixteen years and served as president for two at the pinnacle of his career.

In 1967, at age fifty, he decided to get out of the corporate world. He had purchased fourteen acres of prime real estate in California. He had tried for two years to connect with McMahon, first with the idea of working for a man who could round out his lifelong corporate apprenticeship. However, by the time Phillips ordered his first Dubonnet to

McMahon's second martini, he really intended to carry on to California, to develop his urban acreage. McMahon worked his way through four martinis while Phillips nursed his mild drink. McMahon warned his guest that the Dubonnet would go stale if he didn't drink faster.

"We didn't talk much about business. He preferred to answer my questions about his horses, especially Majestic Prince. I tried to respond with a somewhat embellished account of the three very ordinary horses I had at my seventy-acre hobby farm northwest of Toronto. I was astonished at his warmth and seemingly sincere interest in my modest circumstances." Finally, the conversation turned to Ed Phillips's future. The older man grilled the younger about his career and his plan to become a California real estate squire. Then he said, "We're not advertising any jobs that you'd be interested in. You've been a president of a good-sized company. But I'm going to build two pipelines from the north slope of Alaska right down through Canada into the States, one for gas and, in the same right-of-way, one for oil. We're going to need lots of help if we do it; would that be of interest to you?"

Surprisingly, Ed Phillips was interested. McMahon said, "Okay, I think you should come on out here and join us." There was no mention of a contract, no discussion of salary, and no discussion of titles. Finally he said, "We can't sit here all afternoon getting drunk. Let's go to the office so I can tell Doug [Owen, the president] and D. P. [McDonald, the company's general counsel] that you'll help us build the new pipelines from Alaska for both oil and gas."

Phillips recalls, "We walked the four or five blocks back to Westcoast's headquarters. He took me into the president's, Doug Owen's, office. With Doug was D. P. McDonald. McMahon said to Owen and McDonald, 'This fellow's name is Phillips, he's going to join us to help build these pipelines from up in Alaska.' And he left!

"They asked me my name again, and what did I do. Was I an engineer? No. An accountant? No. A lawyer? No. Well, what can you do? 'I'm a generalist,' I replied. 'I've been the CEO of a fair-sized manufacturing company and I've been in the gas distribution business, some connection with pipelines and so on.'"

McDonald said to Phillips, "Well, I don't know about this pipeline from the north. That's Frank's idea, not ours, but do you know anything about cutting costs?" Phillips replied, "I guess so."

Owen went on, "Well, we're in trouble here, we haven't made any money since we started and there's no way we're going to make any money unless we have some severe cuts in cost and so on because we've built this pipeline on very weak economics."

"I know budgets, finance and cost control. I can help you with that," Phillips replied.

IN RETROSPECT, Ed Phillips's thirty-year career before arriving at Westcoast was the perfect apprenticeship to equip him with the fresh pair of eyes that, with detachment and comprehension, could see what was wrong with the company in crisis and determine what to do about it.

Ed Phillips was a member of the first generation of the Canadian oil patch's Saskatchewan mafia, executives with the common experience of the prairie's empty skies and pitiless winters that bonded them in a common loyalty through the corporate wars. He was born in Saskatoon, as a son of the manse. His father, Dr. Charles Henry Phillips, was successful, though never to be wealthy. The boy's formative years were lived in Los Angeles, where his father presided at a prestigious downtown pulpit. Hollywood seized the young man's imagination. He hawked newspapers so that he could read the tabloid tales of Tinseltown that his mother forbade at home. He hung around the studios and got work as an extra in such extravaganzas as *King of Kings* and *Ben Hur*.

Dr. Phillips, like all Protestant ministers of his era, was a migrant, and his pilgrimage from church to church resulted in Ed's attending eight schools before he was old enough to leave the strictures of his religious home. A talented lacrosse and hockey player and a sometime cowboy, Ed enjoyed a couple of years of rootless, exuberant liberty before arriving in Ontario in 1938 and getting a job with the Loblaw Groceterias Company as a buyer. He lacked a college education but loved the business of business, and was ambitious. He was determined to learn it all and become a generalist able to see the whole picture of whatever company employed him. The key to his success, he believed, would be to work for good executives who would mentor him.

In 1942 war interrupted the careers of an entire generation of young Canadians. Phillips enlisted in the Royal Canadian Air Force. His leadership and his gift for working with all types of men and women were recognized, and he was made a flying examination officer. When Ed Phillips's business career resumed in 1945, he joined the Canada and Dominion Sugar Company in Chatham, Ontario, as its advertising manager in name and, in practice, its minister of external affairs. "I'm a lobbyist," he said to a friend.

As a strategic material during the war and in the economic decompression that followed, sugar was under government control, and Phillips's company was a linchpin of the industry. It was pushed and pulled by the sugar beet growers, the independent refineries and the consumers, and tormented by the uneven hand of national government regulation and international peace politics. As a relatively junior man in the company, he was in fact at the heart of a resource industry that by circumstances rather than ownership had been forced into vertical integration and depended for its survival on what became known, in the 1990s, as strategic alliances. Ed Phillips played a pivotal role to ensure Canada and Dominion Sugar succeeded in that murky, treacherous milieu.

It equipped him for the five years he next spent at Consumers' Gas Company of Toronto, in the executive suite as assistant to Edward Tucker, the general manager. Tucker earned a reputation as an industry-builder, one of a generation of Toronto men who were the architects of the postwar industrialization of the country. At the relatively young age of thirty, Phillips was understudy to a high-octane energy industry architect.

Consumers' was an integrated coal gas manufacturer and distributor—an important domestic utility in the growing city. When Phillips joined the company, the first major American natural gas pipelines were being constructed and the coal gas business was disappearing at about the same pace that crude oil was replacing bulk coal as a furnace fuel. During Phillips's years at Consumers', petroleum displaced coal as Canada's major energy source.

When Phillips arrived in Tucker's office, Consumers' was planning a transition from coal to petroleum fuels. Rather than refurbish and modernize its coal gas plant near the shore of Lake Ontario, close to downtown Toronto, Consumers' wanted to gain gas supply from a major pipeline to be constructed from either Texas or Western Canada. Texas seemed the better bet, as its gas fields were mature and the reserves well established, whereas Alberta's fields were new and the provincial government was opposed to exporting gas until there was a large enough supply for the province's permanent needs. However, Westcoast Transmission's Frank McMahon had gas he'd discovered in northeast British Columbia and was looking for places to sell it, so there were dealings between Consumers' and McMahon.

Consumers' opted to bring gas from Texas through an existing pipeline network that extended from the Lone Star State to Niagara Falls. Consumers' planned to build the connection through Niagara Falls to Ontario, pulling gas into its Toronto market. C. D. Howe thwarted the scheme. He determined that Canada needed to keep its distance from the powerful American petroleum industry; to develop a measure of self-reliance. Howe had seen during the development of the Canol project during the Second World War, and in American handling of the Asian rivalries that soon ripened into the Korean War, that the U.S. would always put its self-interest first, at Canada's expense. Howe used his authority over the Department of Transport to declare the Niagara River a navigable waterway and therefore under his control. He refused to grant permission for a pipeline crossing from the U.S.

Once the plan for a connection to Texas was defeated, Consumers' developed a gas swap plan. McMahon would send gas to the U.S., and Consumers' would pay him and swap the volumes going into the Pacific Northwest for the volumes coming from Texas into Ontario. The financial community knocked down that idea. It refused to believe that McMahon would ever get his line built. C. D. Howe also opposed it in favour of supplying central Canada from the all-Canadian TransCanada PipeLines.

Phillips first heard of Frank McMahon during these events and was fascinated, but he did not meet him. "I remember talking to people in Toronto, like the investment bankers, about this crazy Irish wildcatter with this crazy scheme to raise money to build a pipeline all the way from British Columbia to central Canada and the United States. They said, 'It's impossible, but the guy is persisting and he's drilling wells up there. He's going to find gas, but there is no market for it in Canada, and nowhere for it to go but the United States and very little market even there for it to support a pipeline of that length.' They ridiculed this man: 'He came to us for financing and we wouldn't give him a nickel.'"

After the McMahon-Consumers' contact ended, Ed Phillips kept hearing McMahon's name. His brother, Al Phillips, later the founder of United Directional Drilling, had worked for McMahon on drilling the relief well for Atlantic No. 3. Al described Frank McMahon as a "wonderful guy with Irish determination, and whatever he starts, he finally makes it work."

In 1952, Phillips's career took its next turn when Trane Company of Canada, the biggest air-conditioning and heating-supply company in Canada, hired him as an assistant general manager. He was promoted within a year to vice-president and became a company director in 1954. He accelerated through the ranks until he was made president and CEO in 1966. His fourteen-year success at Trane made him financially independent and rounded out his corporate education.

After he became Trane's president, Phillips decided he wanted to return to the West Coast, perhaps San Diego or Los Angeles. If he had roots, other than in his relationship with his wife, Betty, that's where they were grounded. When he turned fifty, in 1967, he was mature, secure and restless. "I had other things I could do," he recalls. He wanted more time with Betty, and he missed his boyhood California connection. "We'd bought an option on a fourteen-acre, serviced property overlooking Pismo Beach, California, that provided me the option to get into real estate development. My brother, Al, however, suggested I look up McMahon. 'He's a guy you should consider working for,' Al told me." For two years, Ed Phillips had tried to arrange a meeting with McMahon in Vancouver.

When he gave his resignation at Trane in 1968, he contacted a friend, Art Mayne, Westcoast's financial champion at the Royal Bank and now the bank's number two executive, reporting to legendary chairman and CEO Earl McLaughlin. Mayne was a director at Phillips Petroleum, which controlled Pacific Petroleums, which in turn owned 28 percent of Westcoast Transmission. He set up the fateful four-martini luncheon between McMahon and Phillips in the bar at the Bayshore Inn.

ED PHILLIPS HAD ostensibly been hired to help Frank McMahon open a "Panama Canal" of parallel oil and gas pipelines out of the North American Arctic, to bring petroleum from Alaska's Prudhoe Bay and Canada's Mackenzie Delta to major markets. But northern pipelines had to wait. Promoting billion-dollar megaprojects was a luxury Westcoast Transmission could not afford.

Certainly, the company had contracted the Arctic virus. In 1969, Westcoast announced that it intended to build a big-inch natural gas line up the Mackenzie River from its delta to its southern headwaters, where the line would split to link with Westcoast Transmission lines in British Columbia and Alberta Gas Trunk Lines in Alberta. The company also formed Mountain Pacific Pipeline to bring an oil line from Alaska through Yukon and British Columbia, with its destination Chicago. These ideas belonged to Charles Hetherington, however, and not Ed Phillips. In addition to staking a claim to what Hetherington regarded as inevitable future development, he saw the proposals as a stock promotion. They would distract financial observers from Westcoast's chronic fiscal crisis, holding out the promise of a lucrative future. The stratagem failed and the company's interest in the Arctic languished for several years.

Meanwhile, Phillips plunged into the real business of the company: survival. Following his introductory meeting with Doug Owen and D. P. McDonald, and an all-too-brief orientation, he joined the inner circle. The company's most senior executives had become an informal crisis management team, though no one thought to use the term in a company in which financial emergencies were day-to-day routine.

The company squeezed out its first dividend in 1967, but the financial situation was desperate. Its rate of return hovered at 3 percent; to pay its debts, and raise the fresh capital that would allow it to grow, it needed five times that number. As each passing year brought the steady expansion that is part of the character of long-distance natural gas main lines, the cash crunch worsened. The system's capacity had doubled, to 750 million cubic feet per day, since 1957, but its debt load had increased. Ironically, the flush of Westcoast's financial fever had produced a corresponding flush of prosperity in the Peace River country, as evidenced by miles of seismic survey lines slashing through the bush and a plethora of well sites, gathering lines and processing facilities carved into the landscape. Everyone but Westcoast seemed to be doing well by the company's grand vision.

All major Canadian petroleum pipelines launched in the 1950s had similarly rough start-up years. Driven by the imperative of escalating demand for oil and gas, these outfits chewed up vast capital sums in continuous construction. Profits were slow to materialize. Banks, bondholders and shareholders clamoured for a return on their investment and wondered if they were out of their minds to finance the arteries of this new petroleum economy. While other major pipeline projects had achieved soft landings and were turning into cash cows for their investors, Westcoast had overstayed its fiscal adolescence.

What Ed Phillips found, as he went over the numbers, introduced himself to colleagues and toured the operation during his first months at Westcoast, was outside the box of his orderly experience at Trane. It was even worse in detail than the general picture Owen and McDonald painted in their first interview. The pipeline had been in the ground for eleven years and was still in terrible shape financially, on the verge of bankruptcy with plenty of competitors ready to say, "I told you so."

Before he could apply his expertise and look for ways to make a profit, Ed was assigned to help complete two commitments with the potential to drag Westcoast under. The unsuccessful effort to get the Yoyo line working had been abandoned and a replacement line was under construction with equity financing from El Paso. The Pacific Northern Gas system was in its final stage of construction. Crews in the Coast Mountains were welding and testing unburied pipe, but even with savings achieved by dispensing with the formality of burying the line, the project had outspent the capital.

At the same time, the company was completing its head-office building in downtown Vancouver. The uniquely designed cantilevered tower was built on a one-story concrete pedestal containing only a glassed-in lobby and elevator shaft at the base and supported by a graceful web of cables at its top. This was Frank McMahon's personal project: its singular physical appearance wasn't intended as a monument to his maverick qualities as a businessman, but to circumvent city zoning requirements so that it could be the tallest structure in the neigbourhood. The resulting impression, for those approaching the structure for the first time, was of a fortress with a vertical moat of air instead of the traditional horizontal one of water. The style fit the grim mood inside; the company was under siege.

Ed Phillips carved out his portfolio as a senior executive by winning the trust of his colleagues and letting them draw him into their problems. He advanced quickly, being appointed a group vice-president and director after only one year at the company. His initiation ended with the appointment of Kelly Gibson as president in 1970 and his concurrent promotion to vice-president of administration, an overdue portfolio created to match his skills and the contribution he was making as the guy giving the company's affairs shape and structure.

"In 1970, Westcoast's vital signs were barely flickering. There was sombre murmuring among the senior executives, the Royal Bank and the bondholders that a funeral might be imminent," he later recalled. He and Kelly Gibson agreed on the problem: they had to find a way to make a great company on paper into a profitable one.

What bothered Phillips most was that the only profit of any significance Westcoast was recording was a phantom called AFUDC— "allowance for funds used during construction." Westcoast was spending a lot of money each year building new facilities. It was allowed, under accounting rules, to take the interest value of the money it was spending and carry it on its income statement. A new discipline had to be established; more attention needed to be paid to the end game of making a profit, not just finding more gas and building a bigger pipeline or more plants. Those things had to be done, but the focus was changed to the profit objective.

One of his worst times was to be assigned by Gibson to ride herd on Frank McMahon's expense accounts, after the founder became an emeritus chairman with an office and a small budget. "Emeritus!" McMahon once snorted to Phillips. "It sounds like I'm already dead." Gibson wanted McMahon to give up the corporate aircraft, and treated his predecessor with deliberate contempt until McMahon conceded the perk. Caught in the middle, Phillips was saved from a confrontation he didn't want by McMahon's philosophical acceptance of the shabby treatment.

When Phillips's wife, Betty, worried about the toll it was taking on him to act as the buffer between Gibson and the rest of the company, he said, "I'm going to stick it out because a lot of things we're doing have to be done. It's unfortunate they're being done the way they are—so rough on people. When we're finished, we're going to have a good company."

The recovery began immediately and progressed steadily. Budget control, money management and financial forecasting were tightened up. Administrative groups were reorganized, information systems were modernized, and policy and procedure manuals were written. Soon, Westcoast's cost of capital fell below the industry norm and it was building more facilities per dollar than its competitors.

The operating groups reduced costs while upgrading standards, methods and equipment. Training, safety and environmental standards improved. Westcoast began to lead pipelines in engineering performance ratios.

An engineering review removed costly bottlenecks in processing, compression and pipeline loops. A fresh plan, called "guaranteed return on rate base," was implemented to provide record royalties to government, record wellhead prices to producers and record profits to Westcoast. "That," Phillips said, "is a hat trick."

As the company turned around, earnings and cash flow per share, return on assets and capital, stock price and dividends all increased. Financial analysts started to take the company seriously. Employees' compensation increased and, once the layoffs for budget reasons ended, the company's turnover of personnel fell to negligible levels. When it hired new people, it attracted the top applicants in all job categories.

Phillips's soothing style based on patience and tact, his rapport with all ranks of the company and his external political skills were, as much as his talent for corporate organization and money management, the qualities that elevated him to be Kelly Gibson's successor. A year after becoming vice-president of administration, he became executive vice-president and in 1972 was named president. Gibson moved on to the joint chairmanships of Westcoast and Pacific Petroleums, so Phillips still reported to the man he referred to privately, ruefully but with grudging respect, as "The Enforcer."

Phillips knew that he and Gibson had successfully ended the years of financial frailty. The orientation to profit had succeeded; Westcoast could promise its shareholders and financiers steady rates of return at competitive levels for a regulated utility. Phillips understood, however, that consolidation was only half the job; the company still lacked the worldly management sophistication it would need in the changing North American energy sector. And it needed projects that would allow it to grow. Hunkered down behind the mountains, Westcoast Transmission could never amount to more than a sleepy little utility unless it regained some of the vision of its founder. No matter how profitable it was, that wasn't the extent of its potential.

Under the shadow of Kelly Gibson, the company's chairman, President Ed Phillips patiently took control at Westcoast Transmission, not in a single day but through a process in which he used his credibility with people to win full command of the company's resources, and to set its direction.

9

The Conciliator's Brand

*The Foothills Pipe Lines victory, in which Westcoast and Alberta
Gas Trunk Line as fifty-fifty partners were awarded the authorization to
build a $26 billion gas line from Alaska to Chicago, was such
a momentous event that a great deal of Westcoast's history seems to
be measured from that point.*

ED PHILLIPS

In the summer of 1974, Westcoast Transmission returned to its long-standing ambition to participate in Arctic pipelines. The company was a latecomer to the fight for a huge prize. In 1970 two groups competed in a race to win American and Canadian regulatory approval and to assemble the capital to build a natural gas transportation freeway out of the western Arctic. A multinational consortium called the Northwest Project Study Group (later dubbed "Arctic Gas"), that included TransCanada PipeLines, Imperial Oil, Gulf Oil, Shell Canada and the three largest Canadian gas distribution utilities—Consumers' Gas, Union Gas and Northern and Central Gas—was running first in the race. They were opposed by a diminutive Canadian maverick option, the Gas Arctic Study lead by Alberta Gas Trunk Line. In 1972, AGTL had reluctantly married into the Arctic Gas Project, but Westcoast chairman Kelly Gibson knew that Alberta Gas Trunk Line's CEO Bob Blair still wanted an independent Canadian transportation system.

The 1968 discovery of the ten-billion-barrel Prudhoe Bay oil field off Alaska, on the north slope of the Beaufort Sea, triggered the competition. There were several trillion cubic feet of natural gas associated with the oil, and continental North America was short of natural gas. Promising, parallel discoveries in the Canadian Beaufort Sea and Mackenzie Delta created a Canadian dimension to a northern gas transportation system. The prize was huge. The word "megaproject" had been coined in the 1970s to describe multi-billion-dollar oil and gas capital investments. This gas line promised to be the biggest of the bunch. Shortages of oil triggered by the 1973 OPEC embargo created a panic over future energy supply, and the high oil and gas prices that resulted from OPEC's cartel management of prices made those billion-dollar numbers seem both possible and practical. The Trans-Alaska oil pipeline cost $9.3 billion; the estimates for the gas line escalated from a back-of-the envelope $10 billion when first conceived in 1972 to $26 billion when suspended in the early 1980s.

The timelines were typical of pipeline politics. It had taken the better part of five years for the Trans-Alaska oil line to clear regulatory and financing hurdles before construction started in 1973. And that line was an Alaska-only route from Prudhoe Bay due south to the port of Valdez, from which the oil was tankered to Pacific Coast refineries in the United States. There was an enormous volume of natural gas—several trillion cubic feet—associated with the ten billion barrels of oil. The problem was how to get it out of the North.

Westcoast Transmission came late to the game. The Gas Arctic Study project had been under development for almost five years and seemingly had all the advantages of a carefully selected route and design by a team of the world's best pipeline engineers. In 1974, Westcoast chairman Kelly Gibson, however, made contact and proposed a pact with Bob Blair: if AGTL withdrew from Arctic Gas, then Westcoast would partner up with him. On March 27, 1975, Westcoast and AGTL filed the Maple Leaf Project, proposing a pipeline down the

Mackenzie Valley to bring the 5.5 trillion cubic feet of Canadian gas discovered to that date to market via AGTL's and Westcoast's systems. When it was time to take the pipeline forward, the Maple Leaf Project was incorporated as Foothills Pipe Lines, with Westcoast and AGTL as equal partners.

The Arctic Gas consortium was made up of twenty-six of the largest oil companies, gas distributors and pipelines in North America, and these giants were incensed at the regulatory turmoil that ensued. They had just filed applications with Canadian and American regulators and had been confident of success. AGTL's desertion to create the all-Canadian Foothills Pipe Lines alternative threw a political wild card—Canadian nationalism—onto the table, and that put their proposal into a tailspin. Arctic Gas unsuccessfully sought to get the regulators to throw Westcoast and AGTL out of the game.

Gibson and Phillips delegated strategy in the battle to Ron Rutherford, Westcoast's most-seasoned project man. He'd worked for Shawinigan Power in Quebec, the B.C. government and Bechtel Corporation. He built the Inland Natural Gas distribution system for John McMahon before Westcoast's Peter Kutney recruited him. Rutherford became the lead hand in the management of Pacific Northern Gas, and its first president. Ed Phillips got to know him while helping out with PNG. Rutherford was the only Westcoast man with serious qualifications as a project developer; it so happened that no Canadian working for any company could match his talent. It was Rutherford who was first to draw the simple but brilliant conclusion that what would make the Foothills pipeline doable was its connection to the existing Westcoast Transmission and AGTL systems in northern B.C. and Alberta. The Arctic Gas project, without Westcoast and AGTL, would have to duplicate more than 1200 miles of big-inch, high-pressure transmission lines to reach markets.

Rutherford and his AGTL counterparts put together a feasible system, and AGTL's Bob Blair mastered the Canadian political maze. Blair recognized that the regulatory decision to award the pipeline prize might be made in Ottawa and Washington, but the overriding political approvals would be wrestled out of small, mostly aboriginal communities in Canada's two northern territories, which were shucking off the vestiges of colonialism, mentored by Prime Minister Pierre Trudeau. The project, however, needed an American champion, and a linchpin between the Canadian partners and American commercial interests. Westcoast's experience and contacts came into play.

Once before, between 1954 and 1957, Westcoast had forged an American commercial alliance with El Paso Natural Gas for markets to support its transmission system and to get U.S. regulatory approval to export gas to the Pacific Northwest states. Now, needing an American partner for the Foothills project, Ed Phillips turned to El Paso's new owner, Salt Lake City–based Northwest Energy Corp. Northwest's high-flying and ambitious CEO, John McMillian, was decisive and daring—more in the mould of Frank McMahon than any of his Canadian counterparts. He quickly agreed to join the Alcan project, a Canadian-American partnership to pipeline Prudhoe Bay gas following the Alaska Highway into Alberta and B.C. with links to Mackenzie Delta fields. Foothills Pipe Lines would handle the regulatory approvals, financing and construction of the Canadian end, and Northwest Energy would deal with matters in the U.S.

The business arrangement was simple, flexible and fiercely competitive, but it was destined to fail because of the conflicting ambitions of the three partner companies and the clashing personalities and styles of Bob Blair, John McMillian and Kelly Gibson. A complex project under intense scrutiny and facing ruthless competitors and impossible deadlines, herding cats would have been easier. Ed Phillips and Bob Pierce, of AGTL, however, forged the link that held the project together. The deference and polish of the Westcoast man came to its highest and best use combined with AGTL's focus and drive, as the business relationship between the partners held together and the professionals in the trenches got on with the job.

In May 1977, the report of the Berger Royal Commission—formed by the federal government to study northern petroleum development—recommended a ten-year delay in Mackenzie Delta development and the crossing of the Arctic National Wildlife Refuge (needed for the Alcan route). Its terms played a role in putting the brakes on Arctic pipelines, but it hurt Alcan least of all the projects. In July, the National Energy Board rejected the Arctic Gas line and endorsed the Alcan proposal, with some route changes that would make it easier to tie in Delta gas in the future. A parallel project, Polar Gas, designed to bring production from the eastern Arctic archipelago down the west side of Hudson Bay to central Canada, came forward at the same time but did not make it into hearings before its economics collapsed.

In spite of winning the battle over western Arctic Gas pipelines, Alcan/Foothills stalled because its backers could not raise the required billions of capital dollars. Financiers wanted government financial guarantees; not only did the governments refuse, but AGTL did not think they were necessary. Months dragged on with the project ensnared in politics. Although in 1979 President Jimmy Carter and Prime Minister Pierre Trudeau threw more weight behind the project, it could not be disentangled from the web of delay. The National Energy Program undermined international confidence in the stability of Canadian policy-making and the sudden collapse of prices a few months later put the project in limbo. Foothills constructed and operated only the so-called pre-build sections of the line in southern Alberta and Saskatchewan to increase the flow of Canadian gas into the United States. They are still intended as important cross-border ties for the eventual completion of a north-south natural gas pipeline system from Alaska and the Canadian Arctic.

In total, the competing factions spent $180 million and ten years with little to show for their effort. In 1975, Ed Phillips and Ron Rutherford tried to bring an early end to the attrition, on the premise that whatever Arctic pipeline system was approved, its costs and engineering requirements were so high as to require the participation of all major North American gas pipeline companies. The Phillips-Rutherford proposal was to merge Foothills/Alcan with Arctic Gas, creating a single proposal. The idea died on the vine, however, when TransCanada PipeLines, still embittered by AGTL's defection from Arctic Gas, refused to play.

In spite of this disappointment, the Foothills project was a career achievement for Phillips, and it lifted Westcoast Transmission's sights to broader opportunities than could be found on the Pacific Coast. As a comfortable, profitable little regional gas utility, it was underachieving. It was founded as a continental company and a pioneer gas exporter to build the first major Canadian–U.S. pipeline system. With Foothills, Ed Phillips restored a sense of possibility to the company, giving it a future in which it was worth investing its profits.

AS WESTCOAST TRANSMISSION'S president, Ed Phillips ran an oligopoly in which the company was the only buyer, transporter and marketer of British Columbia gas production. He knew that could create lethal rivalries with powerful producers who wanted to cut the company down to a size they could manipulate and control. He knew that oligopolies sometimes bear an impossible financial and social burden to meet the noblesse oblige of their position in the economy.

The oligopoly problem was compounded under the jurisdiction of strong provincial premiers who wanted to enhance the monopoly powers of utilities to control them. One achievement of the Phillips presidency was to cement Westcoast's relationship with the nascent NDP government and Premier Dave Barrett. Kelly Gibson ordered Phillips to get along with the socialists, even though most energy executives hated the government's philosophy and had a bitter tribal memory of the short shrift they'd gotten from Premier Tommy Douglas in Saskatchewan.

Phillips, however, found that Barrett had no particular difficulty with Westcoast's oligopoly, as long as it was "an instrument of public good." It fit well with the premier's notions of a mixed economy and at times gave the government a degree of indirect control of the economy's management that suited Barrett's agenda. Phillips set about finding common cause with Barrett as often as he could. The resulting relationship helped, among other important ways, to counterbalance Pacific Petroleums' conflicting interests as the controlling shareholder of a pipeline and as a producer whose economic advantages sometimes came at the expense of the pipelines that bought its products.

The trap in Pacific's market control was that, as sole purchaser of the natural gas in northeast British Columbia, Westcoast could be forced by arbitration to increase the price it paid for gas supply, while its exclusive customer base was protected by twenty-year, firm-price contracts. The producers applied successfully for three arbitration decisions, each adding to Westcoast's overhead without giving the company any means to recover the cost. El Paso, now the purchaser of 75 percent of Westcoast's gas at the downstream end of the pipeline, volunteered an increase in payments but wanted the Canadian utilities to disregard their contracts and pay the increase too. El Paso wanted to assure Westcoast's survival, but the Canadian utilities refused to join the deal.

Then, San Francisco–based Pacific Gas and Electric volunteered to Alberta producers a substantial increase for the gas it purchased there to ship on the California pipeline it had constructed in 1961. PG&E could afford the generous tender because it owned both the pipeline and the gas aggregator in Alberta that would receive the increase—it was money going from one pocket into another for the huge California utility. The Alberta government promptly ordered that any arbitration in Alberta must result in an award no less than the PG&E family deal.

Imperial Oil, one of Westcoast's fiercest rivals, saw the blood in the water. It asked for another price increase from Westcoast, and forced the matter into arbitration. Meanwhile, Dave Barrett's new NDP government had instituted a very tough public hearing into Westcoast's monopoly privileges with respect to virtually all of British Columbia gas production.

Faced with the "leg trap" of firm price contracts and open-ended gas prices at the wellhead and worried by the vagaries of British Columbia politics, Phillips decided it was time for Westcoast to get out of the gas marketing business. With advisors Ron Rutherford and Jack Smith, Westcoast's chief financial officer, he arranged with Premier Barrett for the province to acquire Westcoast's gas purchase and sale contracts as a business that would be turned over to a new Crown corporation, the BC Petroleum Corporation.

When Imperial Oil challenged for higher prices and went to arbitration, Westcoast simply stayed away from the hearings and Barrett accelerated the negotiations for the transfer of the marketing interests. Imperial won the increase, but Westcoast never paid the hikes because, by then, the deal with the province had been consummated. The deal was brilliant because everybody got something: the province gleaned higher royalties while the producers and the pipeline ended up with assured profitability and much greater stability.

The Phillips–Barrett relationship did not always flow smoothly. During the early 1970s, Amoco Canada Petroleum developed a new natural gas field at Pointed Mountain in the Northwest Territories, just over the B.C. border, and connected it to the north end of the Westcoast system. Pointed Mountain looked to be the most prolific gas field discovered in Western Canada since Turner Valley. Westcoast staked a third of its long-term gas supply on contracts from the field. Then disaster. Virtually overnight the wells "coned," an expression used to describe the engineering phenomenon that takes place when the flow of gas into well bores draws up the water that lies below the gas-water contact line into the gas reservoir. When the Amoco wells at Pointed Mountain watered out, the damage was irreparable. The catastrophe took place just before the winter heating season, and Westcoast announced, *force majeure,* a contractual curtailment of deliveries equally to customers.

The Canadian utilities, particularly BC Hydro and Inland Natural Gas, went to the premier. They painted a drastic picture of economic devastation and job loss. Barrett took the bait, drafted a "B.C. first" policy and ordered Westcoast not to cut supplies to BC Hydro, Inland Natural Gas and Pacific Northern Gas. The company had no choice but to comply: Barrett said he would seize its facilities and lock the valves across the border if they did not.

In the ensuing international incident, the governors of the Pacific northwestern states claimed dislocations of 100,000 jobs over the two winters that it took Westcoast to replace the lost Pointed Mountain gas supply. The U.S. State Department accused Canada of breaching international accords, and Ottawa conducted its own inquiry to placate Washington. The issue faded away because the injured parties had no recourse in an era of regulated pipeline franchises that had monopoly power and plenty of arrogance to match it. This was a case in which the interests of British Columbia and the national were aligned with Westcoast's.

Barrett found the affair instructive. It fed his socialist distrust of business, and with his characteristic relish, he never forgot the incident. Nevertheless, he became one of Westcoast's most helpful political champions, and in 1976 used some of the revenues pouring in from gas royalties to buy the 13 percent stake in the company still owned by El Paso. He wanted to raise capital to acquire the entire company and turn it into a provincial Crown corporation, using BC Hydro as a corporate model. But the voters terminated his political mandate after only one term, leaving Phillips with the diplomatic task of explaining to the incoming Social Credit premier Bill Bennett that he hadn't gotten along so well with the NDP just to embarrass the Socreds.

WESTCOAST TRANSMISSION HAD become a family company knit together by crisis and adversity. Ed and Betty Phillips gladly joined the circle, attending the parties, touring every field location and getting to know folks by their first names. Ed made a study of his people and plunged into every aspect of the company's life. After meeting people, he'd make a note of their names and record details of their families and careers. Then, he'd study these notes before his tours so that when he met the same people again, he'd be able to greet them properly.

He quickly noticed how the atmosphere improved as he got away from the tension of head office. Bernie Guichon once said that sixteen hours in the field equalled eight hours in the Vancouver office because the field was a much happier place to work. But around the office, it was obvious that Phillips loved his work; it was the job of a lifetime. The feeling was infectious. He had the benefit of analytic detachment. He was a student of the company; fascinated and enthralled; he poured energy and passion into his work. His relationship with Kelly Gibson was difficult and the extreme frustration led to a minor heart attack. He did not pass the punishment down, however, and his attentiveness and consideration became the antidote for Gibson's roughshod style.

He was easygoing and informal. Sometimes that was misunderstood. One of his trademark habits was to walk up the stairs every morning to his office, rather than take the elevator. On a particular morning, a young woman he did not know was ahead of him. As he climbed, she seemed to be hurrying faster. Finally she stopped on a landing, backed into the corner and clutched her skirt as he passed by. Later, to his amusement, he heard that the woman, who did not recognize him, had been afraid that he might be a stalker.

Ed Phillips found it great fun to have a close and open rapport with the employees. One of the highlights of his career was to be fired by the company's social club. He'd been appointed for several seasons to perform as Santa Claus at the annual family Christmas parties. He had a suitable kit, including red suit, black boots and a beard. He recalls, "One Christmas I was changing into my suit and I found that I hadn't brought my beard. I sent a message down to Dr. John Sehmer, the company's doctor, who was there with his kids. I asked him to go to his office and get a lot of cotton batting and some glue.

"We made the worst-looking long white beard you'd ever seen and I remember the little kids sitting on my lap and staring at this weird, phony thing. The next year, they thanked me very much for all the years I'd played Santa and told me I must be tired of all this. They told me they'd got someone older and fatter for the job, but that I'd be made Santa Emeritus!"

IN THE FIRST SIX years of his presidency, from 1972 to 1978, Edwin Phillips turned Westcoast Transmisssion from a seat-of-the-pants enterprise into a business with a degree of maturity commensurate with the size, scope and potential profitability of the assets. He rescued the company from its certain fate as a flamboyant, colourful wreck. Westcoast had been destined to become a barroom legend of industry, like Dome Petroleum or PanArctic or Bobby Brown's Home Oil, a story told over whisky and cigars by the raconteurs of the Petroleum Club to embellish long winter nights of poker. Instead, Westcoast Transmission put it all together: profitability with stability, good order and a restored vision. As the great consolidator of the company's fortune, Ed Phillips was remarkable for establishing successful relationships with players of all temperaments and ambitions, from Frank McMahon, who hired him, through Kelly Gibson, who was his taskmaster for half his presidency, to John Anderson, who became Ed Phillips's successor.

Kelly Gibson once accused Phillips of trying to pacify those above and below him and told him it was his fatal weakness. Those above and below him, however, from the Vancouver blueblood directors to the lowliest field hand in Fort Nelson—all of whom Phillips knew by their first names—understood that the boss valued their work. What Gibson labelled toadying was, to others, the exercise of respect and decency, and it infected Westcoast's corporate culture with an enduring sense of decency towards all those who encountered the company and its works. Phillips's way with people was uncanny. More than a talent, it was a leader's gift. It made him the moral centre of the company, as well as its titular leader, and natural successor to the founding father.

With insight and courtesy, Phillips embodied the Westcoast style. In his tenure, the company consolidated its assets and services into a secure franchise. As he accumulated the respect and loyalty of the people around him, he let the men and women working for the company give shape and form to Westcoast's potential. More than most CEOs, Phillips was at ease in letting talented executives around him take initiatives and risks. The organization chart may have placed him above his people, but in practice he treated them as equals, deferring to their judgement and giving credit where due.

Under his leadership the company became the textbook utility of its time, responsible to a fault, quiet and unobtrusive to the point of being dowdy. That it was too cautious, too secure, too complacent and too insular would not be apparent for several more years.

Two men played especially strong roles in Phillips's successful tenure: Jack Smith, his chief financial officer, and Ron Rutherford, the enterprising engineer whose strategic sense placed him at the heart of Westcoast's most important expansion projects. Rutherford and Smith earned their spurs as co-architects of the 1973 gas-pricing plan, which extricated Westcoast from the financial ruin it faced when gas producers sought higher returns through price arbitration, and the pipeline faced the loss of 30 percent of its gas supply (when Amoco Canada's Pointed Mountain gas field in the Northwest Territories failed).

Rutherford, a former Bechtel engineer, joined the Westcoast organization from B.C. Inland Gas in 1970 to become president of the nascent Pacific Northern Gas, putting it on a better footing. He was the architect of the Kitimat methanol project, which fell victim to the provincial government's political machinations, and the designer of a coal slurry pipeline to link British Columbia's northeast coalfields to the coast, an idea that fell victim to railway politics.

It is in the nature of pipelines that many, if not most, strategic projects do not go forward; Rutherford, however, had his successes. His most remarkable success may have been the completion of the Vancouver Island pipeline twenty years after he first designed it, while at Bechtel, and presented it as a project proposal to Frank McMahon. His persistence in pursuit of this idea ranks with McMahon's thirty-year tenacious quest to build the Westcoast Transmission main line. In Ed Phillips's opinion, Rutherford's best moment was his central role in Westcoast winning the Alaska Highway pipeline pre-build pipeline.

While Rutherford was out on the prow of the ship looking into the future, Jack Smith was down in the boiler room making sure that the company had the financial fuel for its dreams. He oversaw the financial discipline necessary to keep the company whole in the early 1970s, making sure that the cure was not worse than the disease. This role put him in frequent conflict with Kelly Gibson, and better than any other executive he learned to manage Gibson's brawling style. As circumstances improved, he ensured that the company's solvency and profitability were not squandered. One of the major responsibilities of a pipeline's chief financial officer in Smith's tenure, when regulated tariffs provided a price-and-profit safety net, was to manage the high level of debt incurred for capital expansion. Smith proved to be a genius at keeping Westcoast's cost of money under control, a skill he most masterfully demonstrated during the regulatory approval and financing of the company's 50 percent interest in Foothills Pipe Lines.

It is remarkable that Ed Phillips so dominates the story of Westcoast's middle years between its troubled adolescence and its years as a flourishing continental powerhouse. During the ten years he served as president, then CEO and finally chairman of the board, he always had a powerful overseeing shareholder: first Phillips Petroleum, which spoke through Kelly Gibson, then Petro-Canada and its powerful chairman and CEO, Bill Hopper. Another man would have stood in the shadows. Ed Phillips learned to lead his leaders.

During most of the years before the unexpected acquisition by Petro-Canada of Phillips Petroleum's controlling interest, Westcoast Transmission was a "West Coast" company. Its management was happy in its Vancouver domicile and content to be a major force in British Columbia's economic life. The larger oil patch east of the Rockies existed only as a phantom of whispers and rumours. The foray into the Foothills pipeline project re-rooted the ambition first planted by Frank McMahon, to be a player in Arctic gas development. Westcoast Transmission, however, needed to be goaded out of its comfortable sinecure.

10

The Transition

Bill Hopper was seen as a buccaneer and opportunist. When Petro-Canada was growing by acquisitions, there wasn't a suite of opportunities to pick and choose from. When they took a run at Husky Oil and missed, the immediate reaction was to buy the next thing on the street, and that happened to be Pacific Petroleums.

BOB FOULKES

With both Pacific Petroleums and Westcoast Transmission profitable in the early 1970s, controlling shareholder Phillips Petroleum reduced its involvement in the affairs of its two Canadian companies to that of a silent partner. When the Bartlesville executives needed capital to develop their North Sea oil discovery at Ekofisk in 1973, they offered to sell out their Canadian holdings for $300 million to the Canadian Development Corporation, an investment vehicle owned by the federal government. CDC was not interested, but the initiative put the companies in play and alerted Ottawa to the opportunity. Kelly Gibson, who had retired as an executive but was still a director of Pacific Petroleums and Westcoast, was tasked by Bartlesville to keep his eye out for the opportunity to deliver Phillips Petroleum what it wanted.

Finding the right buyer took time. Corporate acquisitions were still a cottage industry overshadowed by the drama of wildcat exploration.

However, with the emergence of OPEC as the world's crude oil price-maker and the price spike that followed the 1973 Arab embargo of the United States, the value of production and reserves and the corresponding value of companies escalated.

In Western Canada the opportunity to make elephantine discoveries overnight had all but vanished and all the major pipelines had been built. Corporate growth by acquisition was becoming an attractive proposition. In 1976, Gibson's patience paid off. A motivated buyer for Pacific Petroleums, and the controlling interest in Westcoast Transmission with it, emerged when the federal government created Petro-Canada. Gibson set up a watching brief, ready to make his approach at the right moment.

Ottawa needed Petro-Canada to give it a beachhead in oil and natural gas operations. The federal government had no direct control over oil and gas policy in Western Canada because of provincial resource jurisdiction. But it wanted to influence the huge investments being made to discover new fields and develop the oil sands. It wanted to ensure that some of the investment flowed to projects in federal jurisdiction—which meant the Northwest Territories and the Atlantic offshore. Prime Minister Pierre Trudeau and his cabinet came from a generation of Canadian Liberals profoundly disturbed by the high level of foreign ownership of Canadian business. In the energy sector, that meant American control over companies making key decisions about the future development of Canadian oil and gas reserves, and collecting the dividends. The government wanted to foster Canadian-controlled participation in a sector dominated by American multinationals. One way was to create a national oil company whose size matched the biggest multinationals. And that meant Petro-Canada was looking for takeovers.

There was a political subtext that made it important for Petro-Canada to succeed. Ottawa was a long distance from Western oil fields and, since the 1947 Leduc oil discovery, communities involved in petroleum development had virtually no elected presence in the succession of Liberal governments that set national oil policy. Liberal ministers struggled to understand the Western regional values that the petroleum industry and its culture embraced, and to come to grips with why their political philosophy, so successful elsewhere, was soundly repudiated in Alberta and more politely rejected elsewhere on the Prairies and on the Pacific Coast.

The provinces' ownership of oil and gas brought with it the right to collect royalties. The Trudeau government believed it was not getting its fair share of revenue from the cash bonanza created as oil prices soared. The cabinet was astonished at the sudden fiscal tilt. The amount of money pouring into the industry's coffers and the treasuries of the royalty-collecting provinces upset the careful balances of federal-provincial revenue-sharing and equalization payments that had been nurtured over years to ameliorate jurisdictional controversies, especially concerning the funding of health care and education. The Liberals believed they were being excluded from the party and that petroleum resources were a national asset, whatever the constitutional letter of provincial ownership might read.

Rising prices siphoned petro-dollars from consumers' pockets and fuelled inflation. It drew vast amounts of capital into drilling and the oil sands. The Canadian economy poured its wealth into the West, and resentment grew in other regions with more affinity for government economic intervention and with more representation in the Liberal caucus.

The generation of seasoned oil executives who had been around since the 1947 Leduc discovery, including Kelly Gibson, studied this political porridge, with its ingredients of nationalism, idealism and regionalism coupled with the fear, greed and ambition of the politicians. It was an explosive mix that would force change on an industry already beset by the turmoil of OPEC, inflation and public hostility about the escalating cost of heating homes and driving cars. When Petro-Canada chairman Maurice Strong and president Bill Hopper set up a makeshift office in Calgary's International Hotel, they were greeted by resentment, anger and loathing from the executives of other companies. However, they had the backing of the federal treasury and, as Hopper said, "These guys would do business with Chairman Mao if his money was green."

Kelly Gibson was one of the first in line, because he had a comparatively sophisticated "get along" attitude to government. It was one of the advantages of working his way up from a 50-cents-an-hour Oklahoma roustabout and not acquiring the inbred political prejudices of his middle-class-born, university-trained colleagues who'd forgotten, if they ever knew, what life was like outside their affluent glass towers in downtown Calgary. Gibson believed that Petro-Canada was here to stay and would become a large and important operating concern. When Hopper, who had emerged as the driving force behind the new company, acquired Atlantic Richfield Canada for $342 million in 1977, Gibson knew this was just the opening round of Petro-Canada's shopping spree. He approached Hopper at a social function, dropped the hint and, when Hopper asked about Phillips Petroleum's interest in selling its Canadian companies, set up a meeting between the Petro-Canada chief and Pacific Petroleums president Merrill Rassmussen.

Hopper wanted Petro-Canada to have a public face through gasoline service stations, in addition to a role in conventional oil and gas operations, oil sands and the geographical frontiers. Pacific Petroleums' chain of Pacific 66 retail outlets was of interest, in addition to its oil and gas production and its stake in Westcoast Transmission. In a series of meetings, however, both parties were diffident. Phillips had financed the Ekofisk field and the North Sea was now its biggest piece of business and preoccupation. Even though Canada was of diminishing interest, the Bartlesville men could afford to wait for their price. Hopper was temporarily more interested in Husky Oil, which had large Canadian heavy oil holdings in addition to its chain of gas stations. Months passed while, in the first half of 1978, Petro-Canada engaged in a bitter bidding war with Occidental Petroleum of Los Angeles for Husky Oil that ended, unexpectedly, in June, when Alberta Gas Trunk Line (now NOVA) snapped up control of Husky in a guerrilla-style stock market raid.

A chance meeting in a Toronto hotel lobby between Hopper and Gibson a month after the Husky battle ended led to the re-opening of secret talks between Petro-Canada and Phillips Petroleum. Hopper travelled to San Francisco, Wichita, Washington and other American cities to ensure that his meetings with Phillips chairman Bill Martin would go unnoticed. Petro-Canada soon faced an old competitor, the not-so-discreet Bob Blair of Alberta Gas Trunk, whose interest, and trip to Bartlesville for exploratory talks, sparked a $7 (from $41 to $48) run on Pacific Petroleums' shares in September. Blair, not knowing of Petro-Canada's interest, bided his time too long.

In November 1978, Petro-Canada inked a deal to purchase Phillips's 48 percent stake in Pacific Petroleums for $62 per share in Canadian dollars, totaling $671 million. When the remaining shares were tendered and the legal bills and brokerage commissions were paid, the friendly acquisition of Pacific Petroleums and Westcoast Transmission cost Petro-Canada $1.5 billion. It was the largest takeover in Canadian corporate history to that date, and the purchase price was a staggering sum, even in an industry in which billion-dollar megaprojects had become routine planning exercises.

WHEN PETRO-CANADA acquired Pacific Petroleums, with its 32 percent controlling interest in Westcoast Transmission, the mega-deal rocked the Canadian oil patch like a tremour registering magnitude 8 on the Richter scale—the number designating great earthquakes. When the federal Parliament legislated the creation of Petro-Canada, the opposition Conservatives argued the exercise was futile because it would take one decade, maybe two, for the Crown corporation to amount to anything. In three short years, by combining the Canadian operations of Atlantic Richfield Canada and Pacific Petroleums, Petro-Canada had blossomed as a senior independent producer. It had a chain of service stations, a major natural gas pipeline system and a range of oil and gas production, including a stake in the Syncrude oil sands project.

The first aftershock rattled the windows at Pacific Petroleums. Merrill Rassmussen and his blunt-spoken executive vice-president Al McIntosh walked away rather than going to work for the government. The ripples of anger spread to the streets. Pacific Petroleums' directors, and the Vancouver bluebloods who considered Westcoast

Transmission an essential part of the British Columbia business establishment, had been left out of the loop during takeover talks. They were incensed. When Bill Hopper arrived at the first Westcoast Transmission board meeting after the takeover, Ernie Richardson, a former B.C. Telephone chief executive officer, said, "You know, Mr. Hopper, there's a special committee of the board that's been formed to buy you out." Hopper asked, "Have you decided on a price?" Richardson replied, airily, "When we do, we'll let you know." One of the drivers behind the plot was B.C. premier Bill Bennett, who thought the takeover of the pipeline amounted to intolerable federal poaching on provincial turf. But the offhanded threat from Ernie Richardson was the one and only time Hopper heard of the matter. Hopper didn't say anything further to Richardson and found the incident "kind of amusing." But he also used his control to force changes on the board, bringing in three Petro-Canada directors including himself—the same number as Pacific Petroleums had nominated to assert its 32 percent minority controlling interest. Among the directors who disappeared was Ernie Richardson.

The anger over the Pacific Petroleums' acquisition rippled into the streets. As the Petro-Canada logo went up on 426 Pacific 66 retail service stations and bulk fuel dealerships, and the Taylor Flats refinery at Fort St. John, a bumper sticker appeared that read "I'd rather push this car a mile than fill up at Petro-Canada."

In Vancouver, Ed Phillips weighed his options. He felt he'd been sold out, and that he'd never adjust to working in what he anticipated would be the restrictive environment of a bureaucratic, politically motivated Crown corporation. Goaded by the Vancouver directors, and well aware of Premier Bennett's anger, he was ready to follow Al McIntosh and Merrill Rassmussen out the door. But he wasn't going to make a decision that didn't include John Anderson, his vice-president of legal affairs and corporate counsel. Anderson was the man who had become, during the mid-1970s, his closest confidante and was being groomed to succeed him when he reached 65 years of age and retired in 1982. John and his wife, Marie, had become close personal friends with Ed and

Betty Phillips. As they closeted themselves together day after day and pondered their future, they agreed that they'd stay or leave together.

With his dark-rimmed glasses, prominent jowls and leonine hair, John Anderson was the corporate executive straight from central casting. A quiet, implacable man, his steady demeanour and sound legal mind made him a key player in Ed Phillips's drive to put some polish on Westcoast's once-unsophisticated business style. Anderson entered the oil business fresh out of law school in 1954 and learned the fundamentals of petroleum law with Imperial Oil, which in those days was the Ivy League training ground for the Canadian oil patch. He was a nephew of another John Anderson, a B.C. property developer and financial backer of the McMahon brothers during the family's Turner Valley wildcatting days. The connection led to an offer from Frank McMahon, in 1960, to join the legal department at Pacific Petroleums in Calgary. In 1970, Kelly Gibson transferred him to Vancouver to take over D. P. McDonald's role when McDonald moved to become chairman of Westcoast Petroleum, a calm sinecure that was the prelude to his retirement. Gibson marked John Anderson as a man with a future, though the two clashed frequently because Anderson was unafraid to stand up to Gibson's blustering ways.

Anderson earned his spurs at Westcoast Transmission in 1973, working on the deal that saw the British Columbia government take over Westcoast's gas purchase contracts through the newly created B.C. Petroleum Corporation. Anderson structured the legal framework for the deal, which allowed Westcoast to buy the gas aggregated by BCPC and continue to sell it through its marketing arrangements with Canadian gas distribution utilities and El Paso Natural Gas. The transaction ended the era of low return on investment for Westcoast, and as its profits and share prices rose, so did Anderson's standing in the company. When the B.C. government went on to purchase El Paso's 13.5 percent stake in Westcoast's common shares, Anderson again participated in creating a sound legal framework.

Amid the anger and uncertainty roiling along the halls at Westcoast Transmission in Vancouver following the Petro-Canada takeover, Ed

Phillips and John Anderson became the cooler heads who prevailed and restored common sense to the proceedings. Bill Hopper can recall the hostility he encountered at Pacific Petroleums, with "guys leaning against the door frames of their offices, watching me go by with a look that said, 'You prick.'"

In Vancouver, Anderson's cool disposition and Phillips's sunny "go along to get along" nature created a more positive atmosphere. They decided not to jump to conclusions but to give Petro-Canada a chance. It turned out that the new controlling shareholder was an improvement. "Bill Hopper gave us more freedom to work, and to get out and compete," Ed Phillips recalls. The litmus test was a quiet little subsidiary called Westcoast Petroleum, formed to operate Westcoast's oil pipeline interests, but which had its own minor oil and gas exploration and production activities. Kelly Gibson had never permitted Westcoast Petroleum to do anything that might compete with Pacific Petroleums. It was a situation in which Gibson's dual role as CEO and chairman of both companies created a conflict of interest that Gibson always resolved in Pacific's favour. "They interfered with our oil and gas exploration, often suggesting we defer to the much larger Pacific Petroleums in certain deals," said Ed Phillips. The conflict wasn't limited to oil and gas operations. In regulatory disputes between producers and the pipelines—for example, on tariff-setting issues—Gibson had always backed the producers and often attacked the pipeline he also ran. Bill Hopper let the pipeline go free.

"The conflict perceived by our gas suppliers was much more damaging. They were reluctant to disclose anything about their future development plans for the simple reason it could be leaked to their competitor, Pacific," Phillips recounts. "Petro-Canada had the same conflict potential, but their approach was entirely different. They simply ignored the Westcoast Petroleum operation, judging it to be too insignificant to merit interference. The industry sensed the difference and trusted us not to be a conduit of competitive information to Petro-Canada."

Hopper also indicated he'd support the existing business plan to diversify the company's interests by pursuing large projects similar to

Foothills Pipe Lines' scheme for Arctic gas. Westcoast was chasing the construction of a pipeline spur to Vancouver Island, the development of a methanol plant at Kitimat, British Columbia, utilizing natural gas delivered from the Pacific Northern Gas system, and doing some early studies on a liquefied natural gas port terminal on the northwest coast.

After a few days of intense discussions, the choice between quitting and continuing became obvious: Phillips and Anderson stayed, determined they'd work with Petro-Canada to make the acquisition a success. Within a few months, the decision had been vindicated. "It worked out, and we were both glad we did it," Phillips was to recall.

"I think they realized I wasn't some kind of dangerous Marxist, and that I wanted Petro-Canada to make money. They understood that kind of language," says Bill Hopper.

WHILE ED PHILLIPS and John Anderson deliberated whether to stay at Westcoast Transmission, Bill Hopper considered selling it. But the acquisition of Pacific Petroleums stretched his thinking about what kind of company he wanted Petro-Canada to be. First he had to rethink the idea of owning service stations as his company's flagships when he learned that British Petroleum had been negotiating to purchase Pacific Petroleums' retail outlets. After a long night pondering the decision, he ordered Pacific's executive to break off the talks because building a retail chain and getting into refining made good sense. "If I had said to the government, 'Look, let's build a big integrated oil company,' they'd have likely stopped me in my tracks. But as we made the available acquisitions, we picked up assets that got us into all aspects of the business," he said.

When Hopper took a closer look at Westcoast Transmission—Petro-Canada's first and only major pipeline—he saw that it made money and had a lot of interesting projects in its job jar. Most important, there was a synergy between Pacific Petroleums and Westcoast.

"Westcoast was the pipeline company for northeastern B.C. and Pacific was the largest single producer of gas in the region. So Pacific was Westcoast's largest single customer," Hopper reasoned. It was a good enough business reason to hang onto the pipeline, for an oil executive who, in spite of working for the government, prided himself on his entrepreneurial instincts, flexibility and penchant for profits.

And Hopper liked the cut of Ed Phillips's jib. In the early going, Phillips stood up to the threats and ridicule of his directors and B.C.'s bluebloods, all of whom thought he should bolt. "Ed was cool in the saddle. We got along very well after we agreed that I wouldn't call him Edwin if he didn't call me Wilbert."

The one thing they didn't agree on was Westcoast Petroleum. Hopper said, "Ed, why don't we consolidate this, buy up all the shares and start building a little oil company?" Ed demurred. "We're just fine doing what we're doing now," he told Hopper. So the idea of building up non-pipeline revenues in Westcoast was parked for a few years.

Of all the strong personalities to cross Westcoast Transmission's threshold in its first fifty years, only Frank McMahon matched Bill Hopper for the ability to think outside the box and build a company using bravado and daring when others might want to rely on more cautious and certain tools. Hopper marched through life so aggressively that he seemed to be perpetually leaning into the wind. He was restless, imaginative and tough. As a federal deputy minister, he was engaged in Petro-Canada from the time it was a glimmer in Prime Minister Trudeau's eye in 1974, doing the initial studies on what the company should be, then crafting the legislation that created it and defined its mandate.

He was born and bred to high politics. His father was an agricultural specialist in the Department of External Affairs, and the family lived in several countries. Bill spent his formative years in Washington, D.C., where his father was posted as a Canadian diplomatic official. He started his university studies in economics, and when he realized that it came by the name "the dismal science" honestly, he switched to geology, graduating from George Washington University in 1955, and went to Edmonton in the post-Leduc oil boom to work for Imperial Oil, spending his first summer on a field party in the Rocky Mountains.

In the winter of 1955–56, he was sent to Dawson Creek, B.C. Imperial was exploring a large tract of Pacific Petroleums' exploration acreage in the Peace River country. The deal was that Imperial would get any oil it drilled on the lands, and Pacific would get the gas. Hopper "sat" on several wildcat gas discoveries; as the well-site geologist, he supervised the evaluation of drilling data, making the initial determination of results and directing the technical response to discoveries. "The gas we found contributed to Pacific's gas reserves, to enable them to contemplate a pipeline south. Little did I know that, some twenty-two years later, as CEO of Petro-Canada, we would acquire Pacific Petroleums and their controlling interest in Westcoast Transmission." The same winter, Hopper also met and married his wife, Pat, who worked in Imperial Oil's exploration laboratory in Calgary.

Western Canada may have been an exciting place in those years of wildcatting and big discoveries, but it wasn't big enough for Bill Hopper. He gravitated back to Ottawa, joined the government and got his first lessons in energy politics working for the Diefenbaker government, creating the national petroleum policy envisioned in the Borden Commission's report. The federal government drew a line along the Ottawa Valley and ordered the refineries on the west side of the boundary to stop importing and take their oil supplies from Western Canada. The government had no legal power to enforce the edict, so Hopper found himself jawboning the Canadian subsidiaries of the biggest companies in the world—Imperial, Shell, Gulf Oil—to persuade them to follow the spirit of the policy.

Then he spent a long, nomadic stretch of his career working for the Boston-based Arthur D. Little consulting firm as a globetrotting advisor to countries and companies developing new petroleum fields and national energy policies in various corners of the world. One of his specialties became the development of national oil companies. He returned to Ottawa, as an assistant deputy minister of energy, when the minority Trudeau government of 1972 to 1974 was coming to grips with foreign ownership issues, and developing its response to the OPEC price and supply crisis. The federal government needed his extensive experience as an expert on creating state oil companies because a Crown-owned oil company was at the top of the agenda.

Hopper once said that in Ottawa, 150 people run the country and the rest are on stand-by. Inside the 150, an influential group of ministers and senior public servants were actively working on a Crown corporation to own and operate oil and gas interests, to counterbalance the domination of the oil patch by foreign multinationals. When Home Oil got into financial trouble in 1972 after attempting to break into the Prudhoe Bay oil play—an obligation that far exceeded its financial resources—these federal officials tried unsuccessfully to arrange a federal takeover of Home.

When that mission was aborted, they began to evaluate the formation of a Crown corporation funded by the federal treasury. Hopper wrote the initial studies—and they are notable for their tough and not always encouraging assessments. One warning he gave the government was that the management of the proposed company, and specifically a strong and determined chief executive officer, would take control of policy-making and the state oil company might turn out to be something very different than the legislators intended.

When Petro-Canada was created, Hopper was preparing to return to Boston. He was persuaded to take its presidency after an attempt to recruit a senior private oil executive proved unsuccessful because the federal pay scale—the same as for deputy ministers—proved to be too low. The Ottawa insiders advising the prime minister on the appointment decided that Hopper's private-sector credentials were strong enough to pass muster at the Petroleum Club, and he was not tainted as a political hack. Initially, he reported to Maurice Strong, the chairman of the company who'd been running the United Nations' environmental agency. Hopper took over the twin posts of chairman and president when Strong departed after just a few months.

When Bill Hopper arrived in Calgary on January 1, 1976, to open the company's doors, he did two things: he established an account with a local taxi firm and he started planning corporate acquisitions. Just twenty-two months later, Petro-Canada gained control of Westcoast Transmission. The placid progress of the years in which the company made the McMahon vision of integrated gas development in British

Columbia a profitable success ended the minute Bill Hopper chugged into its boardroom. And yet, after a period of uncertainty about Petro-Canada's motives and methods, and in spite of the background cacophony of political conflict and public controversy over the National Energy Policy, Westcoast's management settled into a cordial and functional relationship with its new controlling shareholder.

The Petro-Canada chairman and CEO did not interfere with Westcoast Transmission's management or mandate. Rather, he slowly opened the door for Westcoast onto a larger world—bringing the prospect of the company becoming a national, as opposed to regional, business force through the expansion and extension of its sleepy subsidiary, Westcoast Petroleum.

Westcoast was never Bill Hopper's preoccupation. Immediately following the Pacific Petroleums acquisition, he fought off a bid by Joe Clark's short-lived Progressive Conservative government in Ottawa, and its Petroleum Club backers, to wind up Petro-Canada, then continued his acquisition strategy by purchasing Petrofina Canada, BP Canada and Gulf Canada. Petro-Canada continued in full flight, accelerating exploration for new fields off Newfoundland and pushing for the commercialization of the Hibernia discovery. It consolidated the eastern Arctic exploration ventures of PanArctic Oil and became an active oil sands promoter. Then, like everyone else, it took the hit on oil prices in the mid-1980s. In spite of his crowded agenda dealing with these matters, Hopper remained on the Westcoast board of directors and actively encouraged Ed Phillips, then John Anderson, to stretch the company's wings.

Ironically, Premier Bill Bennett and the B.C. government became the troublemakers. In spite of his intentions to promote British Columbia business, the premier did not trust Petro-Canada and wouldn't cut Westcoast a break under its ownership. "British Columbia is not for sale," Bennett would say. "Here we have a federal agency controlling the entire natural gas industry in British Columbia and I won't stand for it." Westcoast thought the premier's position was confused. He did not seem to understand that the province, not Ottawa, had jurisdictional control over oil and gas in B.C., nor was Petro-Canada asking Westcoast's management to sell the province's interests down the river.

At first, the premier ordered Westcoast's lawyers to work with government lawyers on a plan under which the province could take Petro-Canada over, or boot it out of the province. But Westcoast feared that open hostilities with the federal government would tempt Ottawa to cancel Westcoast Transmission's gas export permits, and would effectively kill the company. So they told the premier that the only thing he could do was offer to purchase Petro-Canada's holding in Westcoast. Bennett asked Ottawa several times to sell their pipeline shares. The government refused to negotiate and told the premier to talk to Petro-Canada directly. But when Bennett approached Petro-Canada's chairman, Hopper told him that he considered Westcoast a keeper. "I like the company; it's not for sale," Hopper told the disgruntled premier.

When Westcoast, Chieftain Developments of Edmonton and two Japanese companies made a deal to build a methanol plant in Kitimat, Bennett killed it by insisting on a contract with B.C. Petroleum Corporation to pay $3.35 per thousand cubic feet for the feedstock gas. That price was in the stratosphere. Westcoast tried to persuade the Japanese, who had contracted for 70 percent of the plant's output, to increase the price they'd pay for the methanol so that the deal could go forward, but the buyers, understandably, said the proposal was not competitive and refused to change their bargain. The project collapsed. Hopper tried and failed to persuade Bennett and his cabinet that Petro-Canada's ownership and financial backing for Westcoast Transmission would increase its contribution to and importance in the B.C. economy. It was an extreme frustration for Phillips and Anderson to be accused on one hand of working for the federal enemy and on the other to see the expansion and growth of the company torpedoed repeatedly by the provincial government. "I was told," Phillips said, "that Westcoast had to separate itself in some way from Petro-Canada or lose the support of the province for anything we expected to do in British Columbia."

The political decision that rankled the two Westcoast Transmission leaders the most was Premier Bennett's refusal to see the company

The McMahon brothers (from left, John, Frank and George) met in the Hotel Vancouver on October 8, 1957, for the luncheon that marked the arrival of natural gas from northeastern British Columbia. This is a rare photo, as John had no direct involvement in Westcoast Transmission's business.
WESTCOAST ENERGY

facing page, top: In 1965, Westcoast's controlling shareholder, Phillips Petroleum, sent Kelly Gibson in as president to turn the troubled company's fortunes around. He succeeded, but his toughness earned him the nickname "The Enforcer." WESTCOAST ENERGY

facing page, bottom: Gibson's successor, Ed Phillips, served as president from 1972 to 1982. His affable, conciliatory style and his frequent field visits to Westcoast's operations and employees earned him affection and loyalty. He nurtured a family-like atmosphere within the company, creating a positive motivation to succeed. WESTCOAST ENERGY

above: Reserved, decent and competent, lawyer John Anderson served ably as president from 1982 to 1987, when he died suddenly of liver cancer. At the helm while Petro-Canada was the controlling shareholder, Anderson had a cautious style that contrasted with the bold intuitiveness of his chairman, Bill Hopper. WESTCOAST ENERGY

below: A former academic and consummate regulatory navigator, Art Willms became the company's second-most-important leader in the Mike Phelps years, serving as chief operating officer and the corporate anchor.
WESTCOAST ENERGY

above: Seasoned lumber-industry and petroleum financial executive Derek Parkinson reluctantly stood in as Westcoast's temporary president after John Anderson's death and participated in the group that considered the company's long-term growth options.
WESTCOAST ENERGY

above: Bill Hopper, chairman of Westcoast Energy while Petro-Canada controlled it from 1982 to 1992, envisioned building a North American natural gas giant on Frank McMahon's foundations. WESTCOAST ENERGY

above: A key member of Westcoast's inner circle during the hectic development of a profitable North American footprint, chief financial officer Graham Wilson engineered the financings that made growth possible.
WESTCOAST ENERGY

facing page, top: As the quarterback of business development, executive vice-president Michael Stewart (left) oversaw projects in Mexico, Indonesia, Australia and China, and the quest for an Alaska-to-Chicago natural gas pipeline. WESTCOAST ENERGY

facing page, bottom: A former University of Manitoba classmate of CEO Michael Phelps, lawyer David Unruh rounded out the inner circle of senior executives by providing the legal counsel needed to resolve issues and crises and to consummate deals. WESTCOAST ENERGY

above: Bob Reid ran Union Gas after Westcoast acquired it, then succeeded Art Willms as the company's chief operating officer; he became the first president of Duke Energy's Canadian operation in 2002. WESTCOAST ENERGY

Westcoast management's links to Canada's federal and provincial governments were a key part of the company's growth strategy. Lawyer Michael Phelps (left, and with Jean Chrétien, above) was the trailblazer for this strategy, transforming Westcoast from a regional pipeline franchise to a $15 billion Canadian energy giant with assets from Mexico to Alaska and beyond. DIANA MURPHY (*top*); DAVE ROELS PHOTO (*left*)

involved in any way with the construction of a Vancouver Island pipeline. Bennett's obstinacy not only delayed that project by ten years but cost the province a related $750 million fertilizer plant which would have rejuvenated Powell River's local economy. A natural gas link to the Island had been on the province's economic agenda since the completion of Westcoast's pipeline in 1957. The company engaged in three attempts to put together a successful project. In 1964, Bechtel Corporation suggested a lateral from Williams Lake west to a crossing at Campbell River. It failed because the market was still too small. In 1968, Charles Hetherington designed a system also from Williams Lake that included a crossing at Powell River and a loop into Vancouver, to back up the main line in case of a major failure in the Coquihalla Pass. BC Hydro successfully opposed this concept because the Crown power company feared it would lose market share. In 1970, Pacific Northern Gas promoted a line from Powell River to Comox as an extension of its system; that idea, again opposed by BC Hydro, got lost in the political shuffle when Dave Barrett's NDP government came to power in B.C. in 1972.

Bill Bennett revived the Vancouver Island pipeline in 1981. He ordered BC Hydro to undertake the project on a non-competitive basis to sidestep federal regulators and the National Energy Board and diminish Petro-Canada's influence on provincial energy supply through Westcoast and Pacific Petroleums. The proposal was a declaration of war in the bitter feud sparked by the National Energy Program.

Westcoast took its gloves off for the fight with BC Hydro, challenging the purported estimate of $125 million the electric utility had made for the cost of their crossing from the Lower Mainland to Nanaimo. That number looked very attractive against the $350 million that Westcoast proposed to spend on the Powell River link, but Westcoast believed the government-backed project would cost at least $250 million and depended on $500 million of federal aid to complete the pipeline and related distribution system. Westcoast forced the premier to back down from his decision to sponsor a non-competitive project.

In 1983, Bennett called a public hearing to review BC Hydro's and Westcoast's competing proposals, and literally torpedoed a new

Westcoast proposal to construct the Powell River fertilizer plant in partnership with Union Oil of California and the B.C. Resources Corporation. The fertilizer plant, which came with a guaranteed purchase contract from Union Oil for its output, was described by Phillips as the linchpin of a pipeline crossing to Vancouver Island because it would require gas supply volumes that would make the Island pipeline fiscally practicable.

The process produced a standoff that outlasted Bill Bennett's political life. When Bill Vander Zalm succeeded him and appointed Jack Davis as his energy minister, the course changed. The new government ordered BC Hydro out of the pipeline game and tossed the future of the Vancouver Island pipeline back into the private sector.

11

The Steward

Myself, Mike Phelps, Derek Parkinson and Art Willms locked ourselves up in a room at the Bayshore Inn. We discussed a long-range plan for diversification and a future profile for the company fifteen years down the road. We looked at all possible opportunities: manufacturing, communications, even old-age homes. But in the end it was obvious what a great fit the oil and gas business was for a pipeline company—and it was the business we knew best.
JOHN ANDERSON

When John Anderson succeeded Ed Phillips as Westcoast Transmission's president in 1982, Bill Hopper became the company's chairman. The two men had the credentials to be an impressive executive team, but there were differences to be worked out. Hopper's daring and intuitive style was at odds with Anderson's characteristically formal, orthodox and cautious approach to business issues. Anderson was not to be given much time to lead the company and his legacy is one of unrealized dreams.

If Hopper and Anderson could reconcile the combination of their contrasting styles, backgrounds and world views, the company would have stronger processes with which to make the big, long-lasting decisions. If they could not, then their divisions might drag Westcoast back to the fractiousness of the Gibson years. Theirs was a relationship that defined the natural tension that lies between the CEO of a major company and a non-executive chairman who comes from a corporate controlling shareholder.

For the tall, imposing Anderson, the appointment to the presidency held the cachet of a hometown boy made good. He'd grown up in North Vancouver and put himself through law school at the University of B.C. by driving a water taxi. When he returned to Vancouver in 1970 to become Westcoast's general counsel and head up its legal department, he was a popular social figure and well known on the city's golf courses. He returned to his love of the sea and acquired a notoriously unreliable pleasure boat that, said his friends, seemed to be almost constantly in need of repair.

His twelve years in the Vancouver head office included a tour of duty as vice-president of administration, and he used the time to get to know the inner workings of the company. As a confidante of Ed Phillips, Anderson earned a reputation around the company as an executive prepared to fight battles for anyone who needed help getting decisions, even on matters outside the legal department.

He ate lunch in the company cafeteria and developed a network of contacts and a loyal following inside the operation. Yet he kept his distance from employees and imposed a dress code: ties and polished shoes, skirts and stockings. He demanded respect for his office. For years, he employed green ink in his internal correspondence and ordered that no one else use the colour. Notes in green, even without his initials, were treated as orders.

Not surprisingly, when he assumed the presidency, he became a hands-on boss. He combined a command of detail and a perfect memory, and this made him an intimidating taskmaster because he expected people to know the answers to his questions right off the bat. When details weren't forthcoming, and a manager had to return to his office to check a file, Anderson didn't bother to mask his impatience. Not surprisingly, he imposed a tighter discipline on the way business was done.

Ed Phillips admits, "Prior to John becoming president, affairs had been just a little casual. For instance, John was quite a stickler on time-keeping in the office. So, as president he made a habit of going down to the front door when the office opened just to say hello particularly to those who were a few minutes late. Once Art Willms came in fifteen minutes after starting time, so John said, 'Art you should have been here fifteen minutes ago.' And Art replied, 'Why? Did something happen?'"

But Bill Hopper had things on his mind other than the dress code, green ink and keeping office hours. Under his chairmanship, Petro-Canada exercised its period of strongest influence on Westcoast Transmission's business. Hopper directed Anderson to develop a growth strategy. The chairman returned to an option he had unsuccessfully urged on Phillips—the further development of the Westcoast Petroleum exploration and production subsidiary.

Under its scheme, Westcoast Transmission would become in essence half a pipeline company, with the security of revenues from a regulated business, and half an oil and gas operator, with the potential for large profits from successful exploration. Anderson doubted the wisdom of the move. Exploration and production were dangerously political. The National Energy Program was two years old and failing; interest rates and inflation in the oil services sector were racing ahead of increases in oil prices and killing the incentive for new drilling and development. The high prices for oil and gas that underpinned the year-long oil boom were softening.

Nonetheless, he agreed the company needed a strategic rethink. So he and Bill Hopper assembled a team of four executives to do the job. The key players were Anderson and the company's two senior vice-presidents—Art Willms, who oversaw operations, and finance man Derek Parkinson—plus a young man Hopper and Phillips had brought in from Ottawa, named Mike Phelps, who had been given the new job of vice-president strategic planning.

Art Willms was born during the Second World War and grew up on a Mennonite farm near a town called Namaka, thirty miles east of Calgary, which was notable for having four competing grain elevators on its

CPR siding. As a boy, he could look west at night and see the light of the gas flares over the Turner Valley oil field. He became a teacher and subsequently began an academic career in econometrics, the application of mathematics to economics. He wrote a master's thesis on utility regulation that gave him a reputation at a time when regulatory expertise was thin on the ground.

When the National Energy Board began to develop a sophisticated regulatory regime for Canadian gas pipelines, his knowledge was worth a premium, hence the intensive effort Westcoast's executives made to lure him away from a university teaching and research career. As his career advanced, Willms became Westcoast's voice at National Energy Board regulatory hearings. He was also an able executive and developed a firm grasp of pipeline operations and resource development.

Derek Parkinson was a career financial executive who'd been MacMillan Bloedel's chief financial officer; he joined Westcoast after MacBlo was taken over by mining giant Noranda.

Mike Phelps, Hopper's protégé, was a Quebec boy who'd studied law in Winnipeg, obtained a master's degree from the London School of Economics, then gone to Ottawa, where he'd served as an executive assistant and policy advisor to Energy Minister Marc Lalonde.

The four men hammered out a fifteen-year strategic plan after reviewing and rejecting all options for diversification except the development of Westcoast Petroleum. Critics later said that Anderson ignored his reservations about oil and gas exploration and production to give the board of directors the answer it wanted to the question about how to grow the company. The claim that Anderson followed a pattern preordained by his directors, however, is at odds with his own account that the plan was a great fit and involved diversification within the business that Westcoast already knew best.

Company traditionalists, in fact, recognized that Westcoast Transmission was treading back along the path of vertical integration envisioned by Frank McMahon when he twinned Pacific Petroleums and Westcoast Transmission in the 1950s. Anderson launched the plan to expand and develop Westcoast Petroleum at the beginning of 1983.

For three years the results spoke for themselves. In 1982, Westcoast Petroleum produced 2,700 barrels per day of oil equivalent (oil gas and gas liquids). By 1985 the number had doubled twice to 10,000 barrels per day. The oil subsidiary acquired Texas Pacific Oil's Canadian subsidiary in 1984, raising Westcoast Petroleum's asset worth to $2 billion. At the end of 1985, the oil and gas operation was contributing $60 million in cash flow, a third of Westcoast Transmission's revenue after expenses ($187 million). Oil and gas was also contributing 15 percent of net income. The projections indicated that by 1990, Westcoast Petroleum would be generating half of Westcoast Transmission's net income and dividends.

The petroleum operation was struggling with a recession cycle that ran especially deep in British Columbia. Commodity prices had tanked, reaching their nadir in 1986. The pipeline's big growth ideas were stymied by economics and politics. Among the projects to languish were the Vancouver Island natural gas pipeline, the Canadian-Japanese LNG project, and a joint venture with United Oil of California and Chieftain Development to build an ammonia fertilizer plant on Annacis Island in the Fraser River between New Westminster and Delta.

The only good news on the pipeline front was the action of Brian Mulroney's new Progressive Conservative federal government to institute natural gas and gas pipeline deregulation on November 1, 1986, with the first concrete results expected in 1987. Deregulation was expected to make business easier to conduct and to stabilize the gas exports that underpinned Westcoast Transmission's revenue. However, the impact of deregulation on future income was uncertain. It would ensure the predictability that pipelines depend upon, but it would also create competition in areas such as gas marketing. Deregulation might be good business but it came without a guarantee of an upside for income and dividends.

All in all, as 1985 ended, contrasting the gains made in oil and gas production with the frustrations experienced in expanding the pipeline and gas marketing business, the strategy to make Westcoast a pipeline and production powerhouse had apparently paid off.

In the first quarter of 1986, the average world price for a barrel of crude hovered at U.S. $10. Westcoast Transmission's revenues went into free-fall. Earnings per share dropped 25 percent. "The drop in prices took the entire energy industry by surprise," Anderson told shareholders at the spring annual meeting.

Westcoast Transmission's management team initiated a rapid response. They slashed budgets. Westcoast Petroleum cut its exploration budget for the year. They sold some assets to fuel cash flow. They used the cash they had to buy oil and gas production and reserves from weaker competitors who didn't have Westcoast's staying power and had to sell assets to pay debts. The price decline had reduced dramatically the value and selling price of oil and gas in the ground; that created bargains for companies who had the financial resources to stay in the game.

Westcoast Transmission's strategy was premised on a recovery in oil prices in 1988. The company believed that prices had bottomed out. John Anderson described it as building a bridge. "I'm a conservative," he told *Equity Magazine*. "Westcoast still has cash flow, which many less conservative companies don't. Our corporate aim is to maximize return to the shareholder, not growth for growth's sake."

One thing he was not—and that was pessimistic. "Oil prices can turn on a dime," Anderson reassured the people around him. Westcoast had been careful in its expansion. And the emerging deregulated gas market had a great deal of financial potential. Westcoast was already doing well buying and selling gas on the new spot market.

Most important, Westcoast hadn't blown its brains out, fiscally, in its expansion. It was big enough and tough enough to hunker down, to wait for the new deregulated natural gas market to take shape and for oil prices to recover. Through 1986 and into 1987, the company played out its plan and fared better than many of its competitors in both pipelines and production.

On a late Friday afternoon in June 1987, John Anderson dropped in, as the weekend started, on John Sehmer, Westcoast's house physician, and asked for a prescription to treat a chronic stomach ailment that had been giving him pain. After the weekend, he came back; the antibiotics hadn't helped. Dr. Sehmer examined him and detected what he thought might be appendicitis or an abscess. That evening, at the University of British Columbia, an ultrasound test revealed liver cancer.

Anderson entered hospital immediately for treatment. "He was a real fighter," said Dr. Sehmer. But the tumour was far advanced and nothing could be done. Derek Parkinson took over the helm of Westcoast temporarily and put out an announcement that John had taken an early retirement. At the hospital, Ed Phillips took up a vigil with his long-time friend and became the liaison between the Anderson family and the company, advising Parkinson regularly on John's progress.

In August, John Anderson succumbed to the disease, twenty-seven years after walking in the door for his first day of work at Pacific Petroleums' legal department. Vancouver's business community mourned the hometown boy who'd helped to focus a hometown company on a continental future. As a vigorous, energetic and still relatively young man, John Anderson's premature passing left the company unprepared to pick an immediate successor.

At John Anderson's funeral, Ed Phillips gave the eulogy as his final service to the company. Just sixteen months earlier, in April 1986, almost out of public notice, Frank McMahon had died at his Bermuda estate. Although the founding father had been disconnected from Westcoast Transmission's affairs for twenty years, the conjunction of his and Anderson's deaths marked a major divide in Westcoast Transmission's history. The summer of 1987 was one of those highly symbolic moments in a company's affairs that stood for a break with a past that would remain respected and regarded as a foundation, and a step into an as yet unknown future.

For the time being, Derek Parkinson accepted an appointment as interim president. Bill Hopper began to work with the board on the choice of the person to move the company forward in the direction that the team of John Anderson, Art Willms, Derek Parkinson and Mike Phelps had started to shape and structure.

The initiative to balance the company's pipeline franchise with non-regulated oil and gas exploration and production fell victim to the vagaries of pipeline politics and commodity price cycles. Neither the board nor the management was happy with the financial results that flowed from those coincidental impediments.

If Anderson had lived longer, he and Hopper, who were both consummate professionals and seasoned businessmen, might have developed the solid working relationship needed to reach a consensus on their differences and settle on a strategic course to move the company forward.

John Anderson used the time he had to develop Westcoast Petroleum as a lever to move management from the comfortable present to the promising future, through a debate on the profile of the company and a fifteen-year plan to grow the company. Directed by Bill Hopper and the board, Anderson engaged his leadership team in a process of looking outward and into the future.

One of Anderson's contributions to the process was to make sure that lessons from the past were not forgotten, and that old mistakes were not repeated. Westcoast's caution had been earned in the difficult process of developing profitability after the financial chaos of the start-up years. Anderson successfully wove this conservative aspect of the company's personality into the adventurism of Petro-Canada's "Indiana Jones" chairman. After Anderson's death, the resulting fabric stood the company in good stead as the natural gas production, energy export and pipeline transportation sectors moved through a deregulatory process towards continentalism and globalism.

A GREAT COMPANY is fortunate if it has a defining moment, usually early in its history, which gives lasting shape to its corporate identity and purpose and becomes the fountainhead of the wealth it creates and the profits it earns. Westcoast Energy has had the luxury of two epiphanies in its emergence as a leading Canadian-based North American natural gas development engine.

Westcoast Transmission's first vision quest took place at midnight in the 1930s, on the banks of the Pouce Coupe River near the border of British Columbia and Alberta. There, the flame of a wild well sparked

Frank McMahon's dream of a pipeline system for the natural gas fields of northeast British Columbia. From that singular moment, through five patient decades, McMahon and his successors built the Pacific Coast pipeline system and field services business that became the company's financial and cultural bedrock.

The second moment of inspiration took place in 1988, when the company's chairman, Bill Hopper, convinced his board to elevate thirty-nine-year-old lawyer and former political operative Michael Phelps from the boiler room of Westcoast's corporate planning team to become the company's CEO. The board was persuaded that, however young Phelps was and however unlikely his background, he had the qualities to map Westcoast's route over the mountains and onto a continental energy playing field, and the savvy to assemble a team of trailblazers eager to take the journey.

With a mandate to make Westcoast Transmission much more than a sleepy regional utility pipeline, Phelps set an aggressive tone and direction, built a Delta Force–style attack team and assigned the players their objectives. Opportunities clarified quickly and exponential growth ensued as the company attained its continental reach.

The outcome of the decision to make Michael Phelps the company's leader was a period of growth and expansion that altered the company's name to Westcoast Energy and positioned it to thrive in the twenty-first century.

None of that was apparent in 1988 when the board of directors appointed a successor to John Anderson. Although Canadian energy gurus believed the country's natural gas abundance would soon produce a major transcontinental player, no one believed that player would be Westcoast.

12

The Trailblazer and His Team

> *Two roads diverged in a wood, and I*
> *Took the one less travelled by,*
> *And that has made all the difference*
> ROBERT FROST

When Spanish colonists of the sixteenth century decided to establish possessions in what they called the northern mystery, inland from the Florida coast and beyond Mexico, they called the venture, romantically, *entrada*. The reality fell far short of expectation; after two hundred years of bloody and profitless experience, they abandoned the continent, without much regret.

Westcoast Transmission's North American *entrada* started with much lower expectations and in a sombre mood because the point of departure was John Anderson's untimely death. When the board of directors sat down to consider succession, they knew that Westcoast Transmission was at a crossroads. One fork in the road led to continued development of the pipeline base under a traditional style of leadership. There were ancillary oil and gas production interests that might produce an upside, but this was the traditional model of integration initiated by Frank McMahon, forty years previously. It was predictable and safe; it ventured nothing daring, attempted nothing bold and placed severe limits on the company's growth.

The other fork led across the mountains to uncharted territories that, in business terms, were worthy of the ancient mariners' chart notation "THERE BE DRAGONS." The continental journey was longer, riskier and required a leader who would not be reluctant to blaze a personal trail. He would be required to find the right combination of talented and adventurous coureurs de bois to travel with him on a continental journey that would transform the pipeline and its reliable revenue stream into a platform for exponential growth.

John Anderson and his inner circle had achieved success in building up Westcoast Petroleum as a vehicle for growth, even on the rollercoaster ride of 1980s crude oil and natural gas commodity prices. But Anderson, the diligent, staid, dependable shepherd of the dividend, had struggled with the oxymoron of running an aggressive utility.

After him, the company could stick to its knitting. Demand for natural gas in the United States was rising again after a few years during which competing against cheap oil had taken the edge off the market. Westcoast could have stayed on the Pacific side of the mountains and experienced comfortable years expanding its trunk line, drilling up gas reserves in the Peace River country to support expansion, while adjusting to the progressive deregulation of gas transportation, marketing and exports. "The Westcoast Transmission of yesterday was a major line and a couple of small additions. They generated a pretty good income, and if you wanted to sit there as the CEO and do very little, you could have a happy existence," pointed out Westcoast director Bob Wyman.

Surprisingly, even the British Columbia bluebloods on the Vancouver-centric board of directors, who in the past had made it a point of pride to ignore the rest of Canada, wanted to break the mould. "This is a good company; can't we do something with it?" they demanded.

The board's decision about succession was, for all practical purposes, Bill Hopper's to make, and when he did, it was based on his trademark outside-the-box thinking. He broke the mould, plucking Michael Phelps out of strategic planning and installing him in the CEO's chair five or ten years before that normally would have been his due, in terms of age and experience. Hopper endowed him with the mandate to reach into the company's pockets and take some risks, not necessarily for short-term gains.

Bill Hopper recounts that the idea of moving Michael Phelps into the presidency was born when John Anderson succeeded Ed Phillips. Anderson was a steady, honest journeyman—a perfect lawyer-utility man; but his penchant for detail and his aversion to risk held the company back. His presidency provided a useful interregnum that gave the board time to find a long-term new CEO. The directors were pleasantly surprised when Anderson evaluated diversification options and became a champion of Westcoast Petroleum as the logical means to grow, and his premature death robbed the company of a period of consolidation.

Promoting Michael Phelps served notice to the employees, shareholders, financial markets and competitors that Westcoast intended to alter its mandate. It was no longer content to remain a comfortable, if minor, franchise. Instead it would have continental ambitions, because only the continent was big enough to provide significant growth investments.

Under Phelps the company took a new name: Westcoast Energy Inc. But by the time it crossed the threshold of a new century, it had outgrown its appellation. Energy yes, but no longer West Coast.

MICHAEL PHELPS LIVES his life inside a Hugh MacLennan novel. In his quiet but effective way, he is quintessentially Canadian: a gentlemanly, impeccably groomed, fluently bilingual workaholic who eats lunch at his desk and hurries home on weekends to take his family—wife, Joy, and children, Erica, Julia and Lindsay—skiing at their Whistler condo. Like the nation itself, across the span of his fifty-four years, he gravitated West. Born in Montreal, raised and educated in the law in Winnipeg with the obligatory out-of-country graduate degree at the London School of Economics, he took his postgraduate education that counted at the university of Parliament Hill as an aide to federal ministers of justice and energy. When he joined Westcoast, and rose through the ranks, he joined a generation of business peers who saw Canadian opportunity defined in continental and international terms.

He lived, for the first twelve years of his life, on the bridge between Canada's two solitudes, son of an anglophone family living in a francophone village called Shawbridge, forty miles north of Montreal between St. Jerome and Ste.-Adele, on the threshold of the Laurentians. From the age of five, when he started at the village school as the only anglo kid, he has been fluently bilingual. Typical of schools in small Quebecois towns, the Shawbridge school provided a rigorous education with plenty of homework and no-nonsense exams. Michael Phelps's father was employed in the hardware business with an importer and distributor of machinery and tools.

Phelps grew up at the end of Quebec's Duplessis era, during the francophone ascendancy and the first phase of the Quiet Revolution. In the 1960s his father faced a transfer—the choice was Winnipeg or the United States. Decidely a nationalist, Père Phelps chose to stay in Canada. Winnipeg, he believed, would be the commercial centre of Western Canada. The challenge to its supremacy was Calgary, while Vancouver was, as always, a territory unto itself. As Winnipeg loyalist Phelps recounts the tale, OPEC decided the supremacy issue in favour of Calgary. "Calgary," he says, "is Winnipeg with jobs." And it is still Winnipeg that he sentimentally prefers.

Young Phelps attended United College of the University of Manitoba. Political powerhouse Lloyd Axworthy was a young professor there at the time, and part of the local Liberal Party establishment that drew Phelps into its fringe. An arts and humanities institution of about 2400 students, United College was a nurturing environment. It completed his gentle Canadian upbringing, placid, unperturbed. He attended law school at University of Manitoba and graduated at twenty-two, a little young to practise, so he took a master's of law at the London School of Economics. While there, he became a part-time city tour guide. He completed his degree in a year and returned to Winnipeg in 1971 to article in the attorney general's office, intending to do litigation. However, politics intervened.

During the two years he spent with the attorney general's office, Phelps developed connections within the Canadian Bar Association and the federal Department of Justice while two Liberal Westerners, Ron Basford of Vancouver and Otto Lang of Saskatchewan, were the incumbent justice ministers. His former United College professor, Lloyd Axworthy, was now an MP, and Lloyd's brother Tom was emerging as a strong political operative in Prime Minister Pierre Trudeau's inner circle. Although the rules of his justice department job precluded Phelps's direct political involvement, the Winnipeg Liberals marked him as a comer and potential party asset.

When he left the attorney general's office for private practice, Phelps's high-level federal cabinet connections moved in on him. In 1976 he was recruited to work in Ottawa as an aide in the justice minister's office. In a late 1978 cabinet shuffle, Marc Lalonde inherited Phelps along with the justice portfolio. When the Liberals lost the federal election early in 1979, Phelps returned to Winnipeg, intending to pick up the thread of his law career. Before he had time to settle in, however, Marc Lalonde called again.

The Progressive Conservative minority government was foundering; an election was in the wind. Lalonde told Phelps that the Liberals expected to return to power; that he, Lalonde, would be the energy minister and needed Phelps back in Ottawa. Phelps demurred at first, but when the Liberals won the election in February 1980, he agreed to return, but only to help Lalonde with the transition period. The three weeks he promised the new energy minister turned into six months, and the six-month commitment kept renewing itself. The government was at the time developing the National Energy Program to respond to the OPEC-managed crude oil price escalation. Phelps decided his new job was just too much fun and stayed on as Lalonde's chief of staff. "It was the best MBA you could have. What I knew about energy was slight, so it was a hard way to learn on the job." Phelps spent two years in the Department of Energy, devoting most of his time to relations with the energy industry and oil companies. He sat at Lalonde's right hand through the energy crisis and the National Energy Program. Meanwhile, Petro-Canada was in its acquisitions phase, so Phelps spent much time with Bill Hopper to study the company and advise Lalonde, who was its de facto controlling shareholder.

After two years, Phelps wanted to move on. Being a ministerial aide in Ottawa was not a career, and it was time to get out and get on with the practice of law. His best opportunities came from business: Dome Petroleum courted him, as did Albert Cohen, of the Winnipeg-based Sony franchise family. The most attractive offer, however, came from Westcoast Transmission.

It was Hopper who suggested to John Anderson and Ed Phillips that they bring Phelps into the company. "He's a bright guy and a lawyer and has learned a lot about the energy business and oil in particular with Lalonde; he's been good in the planning area. He's not a pipeline engineer and not a technical guy, so planning was a pretty good area to assign him," Hopper recalled. "He had national experience, was well spoken in two languages, had the best Ottawa contacts and knew the political game. He had a CEO's view of the world; he knew how the world was changing and how to cope with those changes."

Phelps got the first call from Ed Phillips. "He and Bill Hopper were persuasive," he recounts. "Winnipeg was home in every sense of the word, and there were worse things in life than to settle back in Winnipeg, but Westcoast was facing a strategic watershed. The company needed to make choices and get on with some seminal decisions."

There was also the issue of succession, with Ed Phillips coming up to retirement. Phelps realized he was being parachuted in, but almost immediately his presence was taken seriously. Westcoast needed his strategic skills, lawyerly style and economic and political depth. The company was struggling through a recession exacerbated by the oil and gas commodity prices slide. Its big, once-promising projects, which were the key to future growth, were floundering.

Westcoast president John Anderson recognized Phelps's special relationship with, and mandate from, Bill Hopper, and brought him quickly into the inner circle that was slowly changing the company's course. In his new job, Phelps became the proponent of Westcoast Petroleum. Hopper specifically wanted Phelps to build Westcoast Petroleum's producing assets through a series of relatively small acquisitions.

Phelps did well in the assignment. His strength in seeking and completing acquisitions was based on his Ottawa-developed skill as a negotiator combined with his contacts around the industry. As he describes it, he orchestrated the disciplines—tax, regulatory, securities, legal, geological, engineering—needed to consummate expansion.

Long before John Anderson's final illness, Michael Phelps had moved the company beyond being merely a pipeline company by taking Westcoast Petroleum in hand personally and building it up through its acquisitions.

NAMING MICHAEL PHELPS chief executive officer left the succession plan only half completed, if Westcoast's new ambitions were to be realized. No one knew better that the new CEO that in high politics and big business only teams get things done. Frank McMahon had his brother George. Ed Phillips had John Anderson. Michael Phelps needed senior vice-president Arthur Willms.

Under the circumstances, Willms, who was on the short list to become president, might have been expected to leave. Hopper, Phelps and the board persuaded him to stay and provide the essential operating genius. Willms knew the pipeline business—he knew how to do the things that Michael didn't know and didn't want to do. Willms became executive vice-president and chief operating officer in tandem with Phelps becoming president and CEO. "The one guy in the organization who has been pure gold for Phelps is Arthur. He became the operational strength of the organization. The combination of Michael Phelps and Arthur Willms was tremendous," said Westcoast Energy director Bob Wyman.

Phelps was seen as young and brash, an animator with energy to burn who wanted to do everything now. Willms was the seasoned veteran, steeped in the logic and lore of the pipeline and its culture. Balancing them as the company's growth protagonist and his operational alter ego was as good as it was going to get. Phelps was the sail, Willms the complementing anchor.

Arthur Willms learned pipelines from the bottom up after joining the company in 1971. A scholar and teacher in his former academic life, Willms became a student of the company in a self-directed course that started with six months of studying National Energy Board (NEB) hearing transcripts. He followed a TransCanada PipeLines expansion hearing from beginning to end, to understand the national regulatory process.

Formerly regulated only by the province except on the matter of gas exports, Westcoast was coming under the jurisdiction of the NEB for tolls and new facilities construction. Willms's baptism of fire, however, came when Peter Lougheed was first elected premier of Alberta in 1971 and immediately launched an inquiry into the prices being paid to Alberta producers for their gas, which he thought were suspiciously low. There were only three buyers, all pipelines: Westcoast Transmission, Alberta and Southern Gas, and TransCanada.

Although he had been with the company less than a year, Willms had to represent the company on his own. He was scared. The night before the hearing, however, he recalled that Winston Churchill overcame his stage fright by reminding himself that the people in the audience were no smarter than he, and knew a lot less about the subject. Willms realized that he knew more than anyone in the hearing room about Westcoast, because he had prepared the evidence. He was on the stand for three days and proved so successful that he became the voice of the company at regulatory hearings for the next twenty-five years. In this role, he mastered the operating detail of natural gas pipelines. He learned the hows and whys of the business from the executive and the field-hand viewpoints; he built a reputation for integrity and earned the loyalty of his colleagues.

As soon as he took charge in 1988, Michael Phelps began to build his team for the future. "Westcoast Petroleum aside, the real changes started in 1988; a new CEO always brings in new people and I needed a team," Phelps says. He started by establishing a decision-making triumvirate of himself, Willms and a financial genius named Graham Wilson, who came from Petro-Canada to become chief financial officer. The trio earned the water-cooler sobriquet "Holy Trinity."

Phelps intended to be the company's strategic architect, so he needed an engineer to bring his drawings to life. He called on Graham Wilson to fill the role.

Wilson was a Scottish-born financial wizard who had studied science at McGill University in Montreal and had gone on to pick up an MBA at the University of Western Ontario, where his academic achievements made him Gold Medalist. He started his business career in the investment business with Richardson Greenshields in Montreal. Then Genstar Development hired him and sent him to Vancouver to help establish a Westcoast presence with the acquisitions of C-Span and Ocean Cement. When Genstar moved its headquarters to San Francisco, Wilson stayed in Vancouver and joined MacMillan Bloedel's finance department. During his ten-year stint with the forestry giant he rose to the position of vice-president finance. He worked with Derek Parkinson and took his place when Parkinson went to Westcoast. In 1983, Bill Hopper lured him to Calgary, where he gained a reputation as a financial mastermind during a five-year stint as vice-president finance and administration of Petro-Canada. In August 1988, the month John Anderson died, Derek Parkinson recommended him to Michael Phelps and, after the two met for a couple of sessions, Wilson joined the company as senior vice-president and chief financial officer, and as the third man in the ruling triumvirate. "When the nickname 'Holy Trinity' got started, I was tagged as the Holy Ghost because I had the lowest profile," he recalls.

The team-building process gained momentum. Phelps recruited a supporting tier for the Trinity. Former Saskatchewan schoolteacher and Independent Petroleum Association executive director Bob Reid came on board. He eventually became president and CEO of Union Gas, then succeeded Art Willms as chief operating officer of Westcoast itself, with Phelps's former classmate and law partner Dave Unruh as senior vice-president law and corporate secretary, and Michael Stewart as executive vice president of business development.

Phelps blended established hands—like Irv Koop (president of field services and pipelines), Al Edgeworth (who later moved to Alliance Pipeline), Ken Retrutiak (senior vice-president) and Ron Maas (senior vice-president of pipelines)—with newcomers such as Joanne McLeod (a lawyer-banker assigned to run treasury), Eric Schwitzer (a merger and acquisitions expert from ScotiaMcLeod), Greg Ebel (of the strategic development group, who was a former assistant to a World Bank alumnus and Deputy Prime Minister Don Mazankowski), Wayne Soper (senior vice president environment), and Bob Foulkes (the communications-vice president, who had spent twenty years at Petro-Canada with Bill Hopper). The group included an inordinate number of self-described "lapsed Mennonites"—Arthur Willms, Irv Koop and David Unruh—as well as a Manitoba Mafia of Phelps, Unruh, Irv Koop, Bohdan Bodnar (human resources chief), Ken Retrutiak, Randy Price (vice-president of tax), and Lindsay Hall (vice-president finance).

Phelps proved to be a wise judge of people, gathering executives around him who were good team players but capable in their own right. When he was done, he had created a younger company, with a depth of talent in terms of knowledge about politics and government and business that exceeded the directors' most-optimistic expectations. As Phelps saw things, "The world of pipelines in Western Canada is aboriginals, trade unions, the United States (because it tends to be about half the market) and local governments. It's all about marshalling external constituencies, and the people who can do that are the people you need around you."

WHEN MICHAEL PHELPS joined Westcoast Transmission in 1982, and again when he leapfrogged into the CEO's corner office, his critics claimed he would never make the grade. His arrival at the company bore a striking resemblance to that of Ed Phillips and Kelly Gibson. All three were generalists and before they came to Westcoast were relative strangers to pipelines. Like Gibson and Phillips, Phelps was parachuted into the ranks by the top executive and controlling shareholder of the day. Phelps had the advantage of inheriting a relatively strong, healthy company and a fully-engaged controlling shareholder with a constructive attitude, so Michael Phelps's impact on the company ranks only with that of founder Frank McMahon.

Before moving to the bridge, Phelps had to prove his mettle in the company's engine room. He proved to be a quick study and a team player. To stay ahead of changes in regulation, markets and commercial values, he drew the maximum economic advantage from Westcoast's portfolio of assets and services. In the first thirteen years of his tenure as CEO, Phelps expanded the company from $600 million in assets to a whopping $15 billion. "Michael is dynamic and has a strong grasp of the major parts of the business and has got a knowledge base which is extremely wide in the general sense and deep on the energy side," said Graham Wilson. "He is a statesman in his ability to interface with the external world and represent the company. He can build that feeling of association and trust which serves the company well."

The team he built around him soon discovered that Phelps was a demanding leader without a lot of tolerance for mistakes. He wanted to know everything that was going on and wanted to be the major decision-maker. Keeping up with him meant matching his energy. Still, he won people over. No matter what the crisis—and it was a rare day without one—he remained entirely focussed on forging a viable and prosperous company. In the various ways he divided his time, applied his power and expended his energy, Westcoast's future was uppermost in his mind. With his personal connections and corporate clout, he could easily exercise more influence than he did. But throwing his weight around was not in character; he preferred results to celebrity.

None of this stops the dreamers from hoping he might eventually go into politics. Senator Jack Austin, who knows Phelps well and shares brain-picking sessions with him, comments: "There was serious talk in Ottawa about trying to convince Michael to go into politics, because he speaks good French, is politically astute and understands economic and financial issues. Today he keeps himself more aware of holistic problems—political and social problems—than virtually any other business leader in Canada. Michael is in touch with big issues, middle-sized issues, small issues and gossip on a daily basis. He's really current. He's ahead of many working politicians on many issues."

Senator Austin believes, "Michael is without question one of the best strategic operational people in Canadian business. He basically came into Westcoast as a business development officer and became the company's chief business strategist. He also became very familiar with its operations, knowing the details of its Mexican investments, for example, as well as the number one guy on the line. One of the nice things about Westcoast under Phelps's leadership is that it has very close, integrated rapport among management members. That is to Michael's credit. He doesn't lord it over his colleagues. He allows them to grow."

His connections know few boundaries. When members of the British Royal family decided to holiday at Whistler, two future kings—Charles and William—chose to stay at his condo. "I work very hard at my friendships," Phelps explains, when discussing his remarkable personal network. "I prefer to do business with friends. Ottawa was a great training school. It was the best MBA you can imagine. Everybody came by, not just people in the energy business, but top bankers, and industrialists—anybody contemplating broad investment strategies. So at the end of the day, I now have twenty of the best-informed people in the country, to whom I can phone to pick their brains. And I don't even have to work that hard to get to them—because they find me."

Phelps's leadership qualities can best be summed in one sentence: He has the rare gift that when he believes in you, you believe in yourself. Which doesn't mean that that he can't be tough. He introduced the "eat what you kill" philosophy to management, and demands results. The turtleneck sweaters he occasionally wears to company functions are deceptive; they should be armour.

One of his pet peeves is cost cutting—or at least, inadequate cost cutting. Westcoast does maintain a corporate jet, but it does not carry a flight attendant. (When the company plane takes off, a dulcet but taped female voice makes all the usual flight announcements. Invariably, someone sarcastically comments, "Great legs.") That and more profound examples of cost cutting are at the core of Phelps's corporate philosophy. "You can't pull the wool over his eyes," remarks Dr. John Sehmer, Westcoast's long-time corporate physician. "He actually reads all those financial statements that parade across his desk and is critical in the right places. He's sharp and a joy to listen to, either in casual conversation or when he's making a speech."

Within eighteen months of becoming Westcoast's CEO, Phelps had moved the company away from its sinecure behind the Rocky Mountains and into the hurly-burly of the continental gas market undergoing a transition, and Westcoast was about to turn the fix-up into a revolution.

WITH TIMING ON HIS SIDE, Michael Phelps took Westcoast's collective organization and jammed its fingers into a light socket. He aimed to create a North American footprint for the company and knew he had leeway from the major shareholder to go out and develop every opportunity on the company's expanding horizon. He had absolute confidence that back in Vancouver there was a team that could deliver the details.

As Westcoast Energy director Bill Neville describes the transition:

> It was a sleepy utility pipeline company dominated by people who were not innovative or exciting. It went through two phases in a relatively short period of time. The decision was made initially to become an oil and gas producer in addition to having the pipeline base, so we built up a decent portfolio of assets. Then in the first eighteen months of Michael's becoming president, with Bill Hopper still in the chair, the decision was made to move out of oil and gas and move into the gas distribution business.

The lever for the breakout was Westcoast Petroleum; the asset that no longer fit the company's future was worth enough to provide the financial platform upon which to diversify in other directions. And for the time being, Phelps needed Westcoast Petroleum as the source of operating cash. Between 1982 and 1987, Westcoast Petroleum had grown impressively. The goal for the parent company was a fifty-fifty split between income from gas transmission and earnings from oil and gas production. Phelps purchased Texas Pacific Oil Canada for $134 million in 1985 from Seagrams. In 1987, he acquired a 53 percent stake in Canadian Roxy Petroleum from Hudson's Bay Oil and Gas for $82 million. In the same year, he bought out AGIP Canada's Western Canadian oil and gas production and properties for $55 million from the Italian government's state oil company, ENI Group. In December 1989, Westcoast Petroleum did its final deal, acquiring a 12 percent stake in Ranger Oil Ltd., a Canadian independent that had blazed a trail as a North Sea operator.

When Michael Phelps became Westcoast Energy's CEO, Westcoast Petroleum was Canada's twentieth-ranked exploration and production company, with financial resources and competitive room to grow. Phelps, however, was shifting gears. There were few attractive deals for oil and gas properties on the horizon. But there were rumours that Central Capital Corp. was preparing to sell its 100 percent stake in Winnipeg-based Inter-City Gas (ICG), which distributed retail natural gas to 440,000 home, industrial and commercial customers in Manitoba, north-central Ontario and parts of Alberta. It also had a propane gas division and a business unit that made and sold gas appliances such as furnaces and water heaters. Phelps called Central Capital Corp. president Peter Cole and asked to be included in the list of interested parties, should the company become available.

This call gave Phelps an inside track, and he had a partner in Petro-Canada, which wanted Inter-City's propane division. The combined offer was Cole's best deal. In the autumn of 1989, Phelps and Hopper announced that Westcoast would purchase Inter-City Gas Corp.'s gas utility and pipeline divisions for $498 million and Petro-Canada would take the propane business for $265 million, while the manufacturing division would be sold separately. The transaction closed in April 1990 and increased Westcoast's asset base by 50 percent, doubling its rate base and number of employees.

Also in 1989, in November, construction commenced on the long-awaited $500 million Vancouver Island natural gas pipeline, with Westcoast a fifty-fifty partner with Alberta Energy Company. With that link came the opportunity to expand the newly acquired gas distribution business.

Westcoast Energy had broken out of the barn, but the race hadn't really started. "The day we bought Inter-City Gas was the beginning of the end for our buying and holding oil companies," Phelps says. "Pipelines and distribution trade on earnings and oil companies trade on cash flow, and they don't marry well as one stock, particularly if you're expected to pay a dividend. After acquiring ICG, Westcoast was primarily a natural gas pipeline and distribution company, and holding Westcoast Petroleums eroded that value. We had to go in one direction."

In the summer of 1992, Petro-Canada—now 20 percent owned by private investors—was assembling the $175 million it needed to go ahead with its 25 percent commitment to the development of the Hibernia oil field off Newfoundland. Meanwhile, its stock price was languishing, hampering plans to sell more of the federal government's remaining 80 percent stake to the public. Petro-Canada had sold a portion of its interest in the Syncrude oil sands project and a raft of small oil and gas properties, but the most obvious big asset that could bring in a large sum quickly was Westcoast Energy. The company's 37 percent stake in Westcoast Energy was contributing $17 million a year in dividends, and Hopper was being pressured by several fund managers with significant holdings to dispose of the Westcoast equity as the best means of funding expansion. Rumours of a pending sale were depressing the trading price of Westcoast's shares.

Meanwhile, a group of directors on Hopper's board had become discontented with the limits of Petro-Canada's relationship as a controlling shareholder of Westcoast Energy. They felt Hopper, who was both chairman of Westcoast and CEO of Petro-Canada, wasn't giving Petro-Canada enough of a window onto Westcoast or enough influence over its affairs. Hopper, however, believed that a 29 percent equity interest didn't entitle him to treat Westcoast as a subsidiary. He had to allow it much more freedom—though he and Phelps were in accord on the big decisions and important directions.

The pressures added up to an inevitable outcome: in August, Hopper announced the sale of Petro-Canada's 21 million shares in Westcoast Energy for $342 million to a group of institutional investors and a retail distribution organized and assembled by ScotiaMcLeod. The Canadian Imperial Bank of Commerce (CIBC) and Graham Wilson played key supporting roles in consummating the transaction.

The common share purchase price was paid in two installments: $8.00 and $8.25. Hopper wasn't completely happy with the $16.25 price; at the time of the sale, the fifty-two-week trading high was $20.00. The proceeds represented badly needed cash for Petro-Canada; his board and Bay Street were happy. "It was fortunate for Michael in the sense that the shares went to widespread ownership, and he didn't have a control block. With no major shareholder, he was free to do his own thing, and his own thing was pretty much what we were doing before the sale," recounted Bill Hopper. In fact, the deal had come not a moment too soon because Phelps was gearing up to take a run at the acquisition of Union Gas and needed some freedom to move. Hopper remained on the board, though Phelps became chairman as well as retaining the CEO title.

With Petro-Canada out of the picture, the government of British Columbia moved to consolidate its relationship with Westcoast Energy, which was one of the most important Canadian-controlled companies headquartered in the province. Through the B.C. Endowment Fund, established in the 1992 March budget, and using Quebec's Caisse de Dépôt et Placement as a model, the NDP government picked up an 8 percent block of Westcoast shares, representing a quarter of the former Petro-Canada stake, for $100 million.

The deal formally buried the hatchet between the company and the provincial cabinet and blotted out all but a faint memory of the hostilities between Ed Phillips and Premier Bill Bennett that had cost the company several good development projects, with B.C. losing the economic benefits. The province considered briefly then rejected the idea of asking for a seat on Westcoast's board, preferring to treat its holding as an institutional-style investment, and ending Westcoast's formal connection to governments.

There was one subplot that briefly unsettled the deal. British Columbia premier Glen Clark wanted to acquire the whole block of Petro-Canada's shares, a move that would have given his NDP government, with its terrible economic and business-relations track record, a powerful hold over Westcoast's prospects. "I was on a boat in the

Mediterranean, off the shore of Sardinia, with my family on a vacation when he reached me," Phelps recounts. "It was awkward, because of the usual cumbersome ship-to-shore radio-phone procedures. But the answer to his question was easy. I said 'no,' and Clark wasn't the easiest guy to say 'no' to."

As a widely held, publicly traded company without a dominating, minority shareholder, Westcoast had enviable freedom to chart its own course. Michael Phelps, trailblazer, used his new-found liberty to buy the sizable Ontario-based gas distribution utility, Union Gas. The North American natural gas industry sat up and took notice.

13

The Eastern Breakout

*Ontario was an obvious market for natural gas. But building the
natural gas system involved more than dollars and cents.
It was to become another chapter in the continuing Canadian dream.*
WILLIAM KAPLAN

When Westcoast Energy acquired Inter-City Gas in 1989 and Union Gas in 1992, it purchased more than a keystone position in Canadian gas distribution, more than assets, cash flow and access to one of the best gene pools of talent in the country. It adopted a piece of Canadian economic history. Westcoast Transmission's story was plenty dramatic and colourful, but it took place on "the other side of the mountains," beyond the ken of most Canadians. Inter-City Gas and Union were key players in one of the core legends of Canada's remarkable transition from a rural and agricultural community to a resource-based industrial powerhouse—the construction of a national natural gas production and supply system.

Inter-City Gas was a direct spinoff of the rush to build the Trans-Canada system, from the Alberta-Saskatchewan border to Toronto, between the summer of 1956 and the autumn of 1958. To finance the massive project, TransCanada needed a network of local distribution companies to build lateral low-pressure delivery lines and sign up for gas purchase contracts. The Byzantine politics that accompanied the project dragged on for six years before construction, but the seminal task of connecting up the markets came together only at the last minute. When Alberta gas began arriving in Manitoba, in the autumn of 1957, a maze of tiny distribution businesses had sprung up like crocuses on the prairie mud; only the Greater Winnipeg Gas system had sufficient size to be viable, and it was owned by Ontario-based Northern and Central Gas.

An enterprising financier, Robert G. Graham, made his career and his fortune by patiently consolidating these local franchises into a steady, profitable private utility. Eventually, he even acquired Greater Winnipeg Gas, through the company-making purchase of Northern and Central Gas, which had its own colourful and controversial background.

Northern and Central began its life as Northern Ontario Natural Gas, created by blue-chip Vancouver oil man and financier Ralph Farris, who lived to regret his involvement because he had been promoted into it by a compromised politician, Phil Kelly, then Ontario's minister of mines.

Farris was the son of John Wallace deBeque Farris, a lawyer and former provincial cabinet minister appointed to the Senate by Mackenzie King. His mother, Evelyn, was on the board of the University of British Columbia for thirty years. An uncle was the chief justice of the province. Later, Ralph's brother John held the same position. Ralph Farris was a member of the Royal Vancouver Yacht Club and raced a 14-metre sloop called *Hawk*. He lived in the exclusive Shaughnessy neighbourhood, drove fast, expensive cars and had a second residence in Toronto's fashionable Lord Simcoe Hotel. Farris attended the University of B.C. and Harvard Business School, worked in New York briefly and came back to Vancouver, where within five years, he had his own seat on the Vancouver Stock Exchange. In the 1950s, he started to spend half his time in Calgary, where he founded Charter Oil, an independent natural gas producer. (In one of history's ironies, his brother's Vancouver law firm would prove to be Westcoast's principal lawyers for forty years.)

In 1954, Farris and another Calgary oilman named Matt Newall incorporated Northern Ontario Natural Gas (NONG) with them was

Gord McLean, a small-time businessman and convicted felon whose cachet was his uncle, Phil Kelly, Ontario's minister of mines and a Northern Ontario booster of the first rank. With McLean essentially fronting for his uncle, NONG promoted a northerly route for the TransCanada pipeline along a right-of-way that gave it better access to mining and pulp and paper communities than TransCanada's preferred route south of Lake Superior. At first TransCanada wanted Northern Ontario to build its own $50 million line from North Bay back into the communities of the northwest. NONG could not raise the capital to do this.

So Farris and his associates had to persuade TransCanada, C. D. Howe and the Board of Transport commissioners to mandate and develop the northern route by assembling a gas distribution franchise that would basically saturate the market. To achieve this goal, Farris hired an executive team of seasoned American pipeliners to assemble a monopoly to distribute gas through northern Ontario. Their objective was to obtain commitments from all municipalities and industrial gas users. They got a lot of help from the newly created Ontario Fuel Board. By spring of 1955, Farris had put together a credible plan, which persuaded nearly every northern municipality and industry to sign gas purchase agreements with northern Ontario. Industrial users were also falling into line.

In early 1956 TransCanada re-evaluated the northern route. As the spring unfolded, the Liberal government decided to provide a bond guarantee to ensure the financing of 682 miles of the northern route from the Manitoba border to Kapuskasing. This was the measure that led to the great pipeline debate, in which Howe invoked closure rather than lose the 1956 construction season. It was one of the decisions that served to bring down the Liberal government and end its twenty-two-year hold on federal political supremacy.

Northern Ontario Natural Gas grew rapidly, fully developing its market in less than ten years, spending $40 million to build its system and maturing into a steady if unspectacular utility in its first years. However, Farris's career was virtually destroyed by a scandal that arose from covertly granting favourably priced share options before construction of the system, to several Ontario mayors and cabinet ministers, including Phil Kelly and, most famously, Sudbury mayor Leo Landreville. The share options appeared to have been granted to ensure that municipalities joined the NONG system and that the Ontario government would cooperate in lobbying Ottawa.

Kelly was banished from Ontario politics and Landreville lost his seat on the Ontario Court of Appeals, because of evasiveness in owning up to the transaction. In a further stroke of historical irony, one of Michael Phelps's first tasks in the Department of Justice in Ottawa was to provide legal advice on how to remove Leo Landreville from the court.

Farris was convicted of perjury at an Ontario Securities Commission hearing and retired to a reclusive life in Vancouver. Northern Ontario Natural Gas, however, prospered and thirty years later had evolved into the Northern and Central Gas group of companies and ICG's prized possession.

When Inter-City Gas took over Northern and Central Gas, it owned private distribution systems in northern Alberta, Manitoba and Northern Ontario. These units became ICG Alberta, ICG Manitoba and ICG Ontario. Later, ICG won the competition for gas distribution on Vancouver Island.

Meanwhile, Inter-City Gas had extended its reach into the lucrative propane distribution business, building a market in Prairie and Canadian Shield towns and farm operations. And it had developed a manufacturing business, based in the United States, which produced and sold natural gas air conditioners, ranges and other household appliances.

As the pièce de résistance of his endeavor, Bob Graham sold Inter-City Gas to Ruben Cohen's Central Capital Corporation. Inter-City Gas came on the market in the mid-1980s not because it wasn't still a sterling asset but because Central Capital had grown too fast and outrun its resources. Lawyer and Westcoast senior executive David Unruh, who was still in private practice in Winnipeg at the time, recalled that Central Capital invited bids for Inter-City Gas, and there were quite a number of utilities and other energy and financial corporations interested in the deal.

UNION GAS WAS STAMPED out of the machine shop of Ontario energy politics before the First World War as the province's manufacturing sector struggled into existence from the swamps and corduroy roads of Upper Canada. It was born in the cut and thrust of the ruthless local politics that produced the Ontario Hydro corporation as the industrial merchants of the region scrambled for markets, the population exploded and the expanding manufacturing complex developed a voracious appetite for energy.

It is a matter of historical record that, in the foot race to find petroleum deposits and commercialize liquid hydrocarbons on the North American continent in the second half of the nineteenth century, Canada won—twice. The first victory came when Dr. Abraham Gesner, New Brunswick's provincial geologist, patented and named kerosene—from the Greek words for wax and oil—by extracting bitumen from tar sands beds in Nova Scotia and distilling it into a smelly, smoky replacement for the world's declining supply of whale oil.

The second victory came in southwestern Ontario, near Enniskillen, when Charles Tripp and James Williams dug—the word designates the hard, sweaty business of shovelling out what amounted to potholes—the continent's first commercial oil wells. In those days before the internal combustion engine, Charles Tripp, an inventor and self-taught mineral geologist and metallurgist, was far more interested in oil as a paving material and petrochemical feedstock than as a fuel—much to Gesner's disgust. Wildcatter James Williams, the more practical of the two, built the continent's first commercial refinery and sold lamp oil out the back door. Geology, however, favoured Colonel Edwin Drake, whose Titusville, Pennsylvania, oil fields held quantities of oil that far surpassed those in the fields of Enniskillen.

Ontario's rolling southwestern countryside produced something far more significant in 1889: the discovery of large natural gas fields. By the end of the century, the blue flame, a product with a ready market to light the street lamps and homes of the burgeoning new Canadian heartland, was making its first millionaires. Gas had the advantage of not requiring processing through the balky crude oil refineries of the day. So gas producers laid cork-sealed wooden pipelines and went into business as the first integrated producing and pipeline enterprises of the world's natural gas industry.

In 1911, three Ontario gas companies—the Volcanic Oil and Gas Company (founded by the fathers of Canada's natural gas industry, Eugene and Denis Coste), the United Fuel Supply Company and the Ridgetown Fuel Supply Company—merged to form the Union Natural Gas Company. These were not humble beginnings. Some companies are born with a silver spoon in their mouths; Union Natural Gas both mined its own silver and crafted its own spoon. It controlled, for the time, vast gas reserves and a distribution system to match. (Coste went on to launch commercial gas distribution systems in Alberta, then returned to Union Gas's board as Ontario's leading natural gas promoter during the Depression.)

After the merger, no one in gas production or sales west of Toronto counted for much against the financial and commercial strength of the new company. The new boys didn't take anything for granted, and defined customer service in the twentieth century. They grew by acquiring municipal coal gasification plants that distilled kerosene from coal. Coal gas supplemented declining natural gas supplies. After the First World War, Union moved from its Niagara Falls domicile to Chatham, Ontario, in the heart of its market area and the centre of the province's producing gas fields. As the Ontario economy grew, Union thrived, steadily expanding its distribution reach to the burgeoning cities of Sarnia, Windsor, London, Waterloo, Guelph and Brantford. The company also nurtured an influence in the development of public policy and regulation that extended to the deregulation era of the mid-1980s and was rivalled only by TransCanada PipeLines.

During the Second World War, Union made the most important discovery in its history, beneath a farmer's field on the toe of Ontario that is formed by the convergence of Lake Huron, Lake St. Clair and Lake Erie. There Union found a group of depleted natural gas fields so tightly capped by the rocks above that they could be used as almost-perfect gas storage vessels. Gas could be pumped down, would be trapped and preserved, and could be extracted later, when required by Union's customers. The place was called, aptly enough, Dawn—a dawn for the fortunes of the company in the second half of the twentieth century. In 1942, Union constructed the first compressors and began the annual cycle of pumping gas into the formations in summer and extracting it, dehydrating it and selling it in the winter. A pipeline was constructed from the American border at Windsor to bring Texas gas to the storage facility. The Dawn facility, which became the largest gas storage hub on the continent, provided Union with a powerful dual business. It could store gas for its own customers and its competitors through the summer, then draw it down for use in the winter.

When the TransCanada PipeLines main line from Western Canada into Toronto was completed in 1958, Union built a 142-mile, 26-inch high-pressure line to Oakville to connect with the TransCanada system and make its storage plant a cornerstone for the entire central Canadian natural gas system.

In the booming 1960s, the Union Gas service web snaked north to the Georgian Bay town of Owen Sound. Union Gas began to evolve towards what it is today—an energy services company, with an attempt in 1968 and 1969 to merge Union and its principal competitor, Consumers' Gas. The deal failed when the two parties could not agree on share exchange values. Then Consumers attempted a hostile takeover, which failed when Union successfully persuaded the Ontario government to pass legislation restricting ownership to 20 percent.

In 1973, Union Gas merged with United Gas of Hamilton, acquiring the steel town's distribution franchise and bringing under its control the last major independent municipal franchise in southwestern Ontario. With the final gap on its distribution map complete, Union Gas turned to the job of getting the most from its market. Over the next two decades it expanded from a $400 million system to one worth $4 billion. Halfway through the growth spurt, it drew the attention of serious money when, in 1984, it flexed the muscles of its ambition and asked its widely distributed shareholder base to approve a seemingly innocuous reorganization. Under the proposal a new holding company would own Union Gas, the regulated gas distribution franchise, and also start up other unregulated business, including natural gas exploration and production in Western Canada.

Union Gas was, by this time, an integral building block in the Ontario economy. Although it was not a Toronto company like TransCanada PipeLines or Consumers' Gas, it was part of the staid Tory-blue Bay Street establishment. It had a a former Ontario provincial treasurer, Darcy McKeough, as its president. The largest single share block, 17 percent, was owned by a consortium called the Great Lakes Group, which included Brascan Ltd., Merrill Lynch, the Canadian Imperial Bank of Commerce, National Victoria and Grey Trustco, and Hees International Corp.

Darcy McKeough set a new course for the company, building out from the regulated gas business into non-regulated business up and down the energy chain. Union invested in Alberta gas production with Hudson's Bay Oil and Gas and Numac Energy. The company vision was to be a leader in discovery and distribution. The Ontario government wanted the company's gas distribution companies protected from the high risks of natural gas exploration, and to prevent utility customers' bill payments from subsidizing the unregulated drilling and development of gas reserves. At the behest of the regulators, McKeough proposed the reorganization, setting up a holding company called Union Enterprises that owned Union Gas and an independent natural gas exploration and production operation called Union Shield Resources.

When Union's shareholders and the Ontario government approved the creation of Union Enterprises, an investment powerhouse coincidentally named Unicorp Canada Corporation, led by financial veteran George Mann and his young protégé, Jim Leech, was ready to pounce. In a series of three transactions undertaken on its behalf by Gordon Capital, Unicorp spent $73 million in cash to buy 4.5 million shares of Union Enterprises on the open market, then picked up Great Lakes Group's stake in exchange for a piece of Unicorp.

Now owning 30 percent of Union Enterprises, Unicorp began a hostile takeover of the balance of the company. When Unicorp's stake reached 49 percent, Union Enterprises retaliated with a share exchange purchase of Calgary-based Burns Food. By the time the Burns Food purchase was completed, Unicorp owned 60 percent of Union Enterprises, but the Burns deal diluted it back to 48 percent.

In spite of the standoff, Unicorp acquired a block of seats on the Union Enterprises board and in August 1985, the Ontario Energy Board, after a long and acrimonious hearing, allowed Unicorp to keep its controlling interest in the natural gas utility, effectively granting Unicorp the victory. In less than a year thereafter, Unicorp acquired 100 percent of Union Enterprises.

The core gas distribution franchise flourished, but the expansion into non-regulated business faltered and George Mann ran into a financial crunch. Union Gas went into play in 1991 as a takeover prospect.

NO ONE IN CANADIAN energy circles was surprised when Inter-City Gas and Union Gas tumbled to an acquisitor. They were stunned, however, that it was Westcoast Energy that swept up the prizes. No one expected the "mousy" little utility from the "wet" coast, led by that supposed "bureaucrat" from Ottawa and his "loose-cannon" chairman, to deal at this level.

The word went out—Michael Phelps is a comer and Bill Hopper is still in the league.

What made the first deal—the Inter-City Gas purchase—possible, recalls David Unruh, Westcoast's senior vice-president of law and corporate secretary, was that Westcoast wanted ICG's gas distribution and Petro-Canada, its former dominant shareholder, wanted the propane division. Unruh, then practising law in Winnipeg, sat at the negotiating table for Petro-Canada and was later recruited by Westcoast.

The architect of the deal was Westcoast's Graham Wilson. He recalls that ICG, "had the four companies, oil and gas production, the gas distribution utilities, propane and manufacturing. All four companies were put up for sale. What they had hoped to do was sell the company as one unit but when you looked at the components of it, it was very hard to see how anybody was going to be interested in all the parts."

Wilson quickly realized that most of the people who had expressed interest in the company weren't terribly interested in the oil and gas production, so they sold that separately to Mark Resources. Westcoast was interested in the utility division and not the other parts but recognized that it would be an advantage to bid for more than just one unit, "because we'd be able to get a better price."

Wilson had come to Westcoast from Petro-Canada, so he went back to his old colleagues and said, "Let's get together on this." They quickly came to the conclusion that neither company wanted the furnace manufacturing business. Westcoast could buy the utilities and propane businesses and flip the propane business to Petro-Canada.

"I'm not sure that two other organizations could have operated quite that way. It was just the coincidence of the common shareholding and also the association of individual people. In any event, the two of us ended up being the ranking bid and having the greatest offering to ICG," Wilson said.

Wilson didn't have the money to put into this sort of deal, so although Westcoast knew Petro-Canada was ready and able to put up its portion, "we had to worry about the portion we were going to pay." Wilson created a financial structure that was fundamentally an uphill loan, allowing Westcoast to leverage the deal while creating an ability to finance it. He went initially to one bank, the CIBC, and to a senior executive named Ken Davidson—"one of the cleverest bankers that I have ever had the privilege of working with," he said.

"This was a relatively complicated structure, particularly as it had to meet tax and regulatory concerns," Wilson recounted. "It wasn't just a simple loan. Ken Davidson and I hatched up this structure that effectively was a loan to a holding company. The money was then loaned to Westcoast, because we had to make sure the tax deductibility was in the right place, effectively without recourse to Westcoast. The reason it had to be done in that manner was to be sure that the regulatory agencies that had oversight of Westcoast were not concerned about it and also that the bond rating agencies were not concerned about it."

Wilson's next problem was a tight timetable in terms of putting the bid in and coming up with the funding. The CIBC was there for its portion of $175 million in a total loan of $350 million. Westcoast needed to bring in two or three other banks for the rest and was running out of time.

One Wednesday night Wilson phoned Bill Minnear, a senior banker he knew at the Bank of Nova Scotia. He said, "Bill we need to have you come in to support this and I'm going to be hopping on a plane to come to Toronto to walk you through the whole deal."

Minnear suggested he meet with somebody at the CIBC to understand their thinking. Wilson referred him to Ken Davidson. Immediately, Minnear and a couple of his colleagues got together with Ken Davidson. Wilson was still booked to fly to Toronto that Thursday afternoon.

"A couple of hours before my flight, Minnear phoned and said, 'You've got the deal.'" Minnear said their objective after Wilson made the first phone call was to turn around that loan application in twenty-four hours; they turned it around in twenty-five hours. He said they missed the window that they had set for themselves.

The next exercise for Wilson was to mitigate the tax and credit issues. "A couple of us were sitting around over drinks in the Toronto L'Hotel. I said, 'Why don't we try doing a plan of arrangement as opposed to the simple acquisition? We can allocate proceeds to increase the value associated with the businesses we're buying and we'll be able to sell them with as little tax consequences.'"

Bob Graham and his chief financial officer Peter Marriot at ICG agreed, so at the end of the day Wilson got a favourable tax ruling that cleared the deal.

"It was done in such a reasonable and efficient manner that I ended up hiring their tax manager, who was instrumental in taking the deal through Ottawa—that fellow Randy Price," Wilson said. Westcoast had spent, effectively, $1 billion to purchase Inter-City Gas, including assumed debt. There were several opportunities to buy more distribution—the most attractive being Union Gas of Chatham, Ontario. Westcoast didn't have the money to go after Union. But there was one asset—Westcoast Petroleum, its exploration and production holding—

it could sell that would both solve the cash crunch and make the company a pure pipeline play for investors who wanted Westcoast to become just that.

Union was a rich prize. It distributed gas to residential, commercial and industrial companies in southwestern Ontario. It also transported and stored gas for utilities and industrial customers in Ontario, Quebec and the central and eastern United States. Like Westcoast Transmission, it had for years been a sleepy utility, but it was an essential cog in the central Canadian gas market. The one thing the Ontario government didn't want was for Union to fall into TransCanada Pipe-Lines' hands—that would concentrate TransCanada's market power far too much.

Westcoast, with no presence in pipelines east of the Rocky Mountains and a growing reputation as a well-run acquisitor, was a reasonable alternative. "We should never have gotten Union," recalls Arthur Willms. "When the opportunity first arose we didn't have the money. We approached them in the summer of 1992 and said that if they were still available in the fall we might be able to do something. Enbridge Energy (formerly Inter Provincial Pipelines) or someone like that could have scooped them up. But everyone else moved too slowly."

Meanwhile, some American companies were becoming interested, and Michael Phelps rolled into high gear, working his Ottawa and Queen's Park connections to keep foreign bidders out. "Michael is plugged in everywhere right across the country where it counts," says Westcoast Energy director Bob Wyman. "That time it counted as never before."

To raise the money to purchase Union, Westcoast did the inevitable and sold off Westcoast Petroleum. In a transaction timed to coordinate with Petro-Canada's divestiture, and with the Canadian Imperial Bank of Commerce as the architect of the complex manoeuvre, Westcoast Energy divested its holdings in Westcoast Petroleum and used the proceeds to lever its purchase of Union Gas.

It wasn't an easy sale. An American investment banker received the initial mandate to put together a deal for Westcoast Petroleum, but nothing much turned up. A short time later, Michael Phelps was fishing in the Bahamas with his friend from Gordon Capital, Bob Fung. In the middle of contending with landing dolphin, barracuda and tuna, Fung asked, "Why not give me a shot at it? I have contacts in Hong Kong anxious to keep pace with Li Ka Sheng [who had just bought Husky Oil]."

"I'll give you three weeks," said Phelps. Three days later, Fung found a buyer and the key to financing the Union Gas purchase.

In October, after evaluating the company, and after a period of negotiations, Westcoast made the formal offer to purchase the Union Gas shares. It closed the deal in November 1992. The price was $16 cash or one common share of Westcoast Energy. The total cost was $618 million, $381 million in cash and $237 million in common shares. Said Phelps: "The Union Gas purchase was a function of us getting at it quicker than any Canadian contender—then blocking out foreign competition."

"The big step for Westcoast outside the pipeline business was building an oil company," recounts Mike Phelps. Then in the process of buying the Inter-City Gas group, Westcoast realized that gas distribution was an attractive way to diversify. "When Union came on the market," said Phelps, "we realized how big it was—a $600 million package—but that we could grow it. And it has tripled in value since we bought it. Selling Westcoast Petroleum and swinging into gas distribution has been a tremendously profitable gain for our shareholders over eight years."

WHEN WESTCOAST ENERGY acquired Union Gas, most analysts thought that it should have been the other way around. Union had emerged as the biggest natural gas distribution enterprise in Canada outside of Quebec. Consumers' Gas of Toronto had more customers, but Union had more assets and a broader-based business with greater expandability. The Dawn storage facility accounted for some of its advantage, but Union Gas also had a cross-border connection to the United States.

Combined, the interconnects to half a dozen important American pipelines and the Dawn facility made Union Gas one of the fastest-growing natural gas market hubs in North America. However, the opportunity had to be worked and worked hard. Westcoast found itself in the customer service business.

For Westcoast Energy, Union Gas promised to be a steady source of profit, but it also offered a laboratory in customer focus and customer service. The expansion of the North American pipeline system had produced the prospect that filling one pipeline could drain its competitor, a redundancy that regulators in the past had prevented but that open markets now encouraged. Instinctively offering intensive customer service seemed to be the way pipelines and distribution companies could gain competitive advantage over their rivals in the world of deregulated natural gas marketing.

First, however, the two companies had to learn to get along. Two strong family cultures had to find common ground. It was as if there had been a marriage in which the bride's and the groom's families both thought it was their child who had married down.

The Union Gas headquarters in Chatham, Ontario, had developed into a fortress against the outside world. While business in the field tended to be micro-managed from head office, the system worked well because head office was lightly staffed, so the system didn't get bureaucratic—the connection between senior management and junior personnel was quite direct, amicable and personal, and everything was on a first-name basis. The relationships were so close that an instinctive hostility to outsiders developed. Even Toronto was disdained as an upstart pretender to business supremacy.

Phelps assigned the personable and politically adept Bob Reid from Westcoast Energy International to become the president of Union Gas. Reid lured the Union Gas family into Westcoast by the force of his disarming personality, and by making his presence felt in every corner of Union Gas's world.

In 1998, Westcoast reorganized its gas distribution group of companies, selling Centra Gas Manitoba and rolling Centra Gas Ontario into

Union Gas. Although on the map Union still presided over a monopoly, it knew that its potential for growth would be in areas of open competition and that regulators were prepared to break up old utility cartels. Competition for long-distance telephone service had allowed new service providers onto lines built and operated by old monopolies. Similarly, independent gas brokers were selling supply contracts door-to-door.

The good news, for Union, was that natural gas transactions in its market hub were growing at a rate of 80 percent per year. The Dawn storage facility had expanded to use sixteen depleted gas fields throughout Lambton County and was handling 1.1 trillion cubic feet of natural gas a year. While 90 percent of the gas originated in Western Canada, more was coming up through the American pipeline system. The North American gas transportation system had, for fifty years, flowed from west to east; in the 1990s, with deregulation and the creation of commodity trading, natural gas now flowed in all directions: clockwise and counter-clockwise. And the use of gas had expanded so that there were two peak seasons—summer air conditioning as well as winter heating. Through complex swaps and trades, a molecule of gas produced at Fort Nelson, B.C., in January could end up, theoretically at least, being stored at Dawn for the winter and then paid for by a householder in Florida using gas in his air-conditioning unit in July.

Under Westcoast Energy's aegis, Union Gas shook off the image of a venerable, Upper Canadian family compact. Its ambition was to become a multi-dimensional natural gas and electricity energy company. The targets of opportunity were expanded to include gas storage and distribution in the Great Lakes states, more natural gas–fuelled electrical generating capacity and the purchase of local electrical distribution systems, as the province of Ontario deregulated the electricity market and if its municipalities sold their power lines.

As a unit, Westcoast and Union were on the inside track to exploit the completion of the Alliance Pipeline to Chicago and the Vector connection to Ontario, bringing new gas supplies to its franchise areas. Its comprehensive web of interconnects to continental Canadian and American pipelines put it at the heart of the best-served natural gas transportation, storage and customer services hub on the continent—on a par with the coveted Midwest-Chicago market.

Although not the biggest, Union was the most-diversified gas distribution utility in Canada and cracked the juiciest natural gas growth market in the country. Union Gas owned the gas franchise in the rich industrial heartland of southwestern Ontario from Windsor to the Toronto City Gate, including a huge chunk of Lake Ontario's economic Golden Horseshoe, from Welland through Hamilton, Burlington and Oakville. Added to its Centra Gas operations, which incorporated northern and central Ontario along TransCanada PipeLines' corridor, plus the North Bay to Sault Ste. Marie mining belt, Westcoast Energy now controlled an Ontario customer base of more than one million residents, commercial buildings and large industrial plants. It had everything that central Canada's engine room had, except for Toronto and the St. Lawrence Seaway.

Union and Westcoast emerged from their merger wiser, feistier and better prepared for the future. Rising competitive forces were sweeping away their monopoly privileges and bringing them face to face with customers and with competitors in the battered flea market of the competitive natural gas market. This is the way that mergers and acquisitions are supposed to work but seldom do: the combination is larger than the sum of its parts.

14

The Agenda

The ultimate creative act of business is intuition. You make intuitive leaps about where you are going, and why.
MICHAEL PHELPS

Coming out of the Union Gas acquisition in early 1993, Westcoast Energy caught a wave of opportunity as demand for natural gas, and enthusiasm for its potential as a fuel of choice, swept North America. As concern over air quality and climate warming escalated, gas began to displace coal and oil at industrial burner tips and for power generation. Other factors fed the fire: gas was an economical fuel and Americans were worrying about their dependence on imported crude oil. Westcoast's battle-hardened leadership was reinforced from Inter-City Gas and Union Gas. The company had gained the financial capability to do bigger deals. The Canada–U.S. Free Trade Agreement had realigned continental business on a north-south axis. And Westcoast had caught a vision—still inchoate but remarkably powerful—of a North American energy business with international connections.

To capitalize on its positioning and opportunities, Westcoast set aside mergers and acquisitions for the time being and became a project company. It had been ten years since the Foothills pipeline battle over the right to ship Alaska gas, but Westcoast segued into a new round of strategic business development at Mach speed, putting out feelers in Asia, Australia, Europe, Latin America and Mexico even before Mexico joined NAFTA, the North American Free Trade Agreement.

Meanwhile, between 1988 and 1991, there was a long-unfinished piece of business to take care of—the Vancouver Island pipeline.

AS PART OF HIS grand scheme to deliver natural gas to every nook and cranny of British Columbia, Frank McMahon coveted Vancouver Island's oil-captive residential and industrial fuel market. In his years at Westcoast Transmission, he shared his ambition to lay an undersea pipeline from somewhere on Westcoast's main line across the Strait of Georgia to Vancouver Island with Steve Bechtel, the head of San Francisco–based Bechtel Engineering, which had helped assemble the market and build the original main line. Bechtel's grasp of West Coast energy use had been pivotal to the development of the Trans Mountain oil pipeline from Edmonton to Vancouver.

In 1961, Westcoast faced a competing proposal—the Magnum Project. Cascade Natural Gas of Seattle and Northern and Central Gas of Toronto floated the idea of a low-pressure, flexible pipeline through the San Juan Islands. The technology, however, failed during tests.

In 1964, McMahon and Bechtel undertook a feasibility study of a conventional undersea line. Called the Island and Coastal Pipeline, it brought a young pipeline project gypsy named Ron Rutherford to the problem. The study proposed that a lateral be built from Westcoast's main line at Williams Lake to the coast and across the strait to Vancouver Island north of Campbell River. A student engineer named R.J. Brown found and recommended the crossing from Powell River to Comox that was finally built in 1990. Brown went on to achieve fame as one of the greatest project engineers in petroleum history and was co-founder of global giant Brown & Root, but his precocious engineering idea languished. The cost of crossing the demanding undersea terrain, the small domestic market and the resistance of Island industries to switch fuel combined to ruin the economics of the project. What

Westcoast gained from the experience, besides Brown's discovery of a way to cross the Inside Passage, was Rutherford's stubborn enthusiasm for the project.

In its most difficult years, Westcoast never had a better friend than Ron Rutherford, a Montrealer, former Bechtel employee and jazz aficionado who once led his own band. Rutherford was the quintessential project engineer and consultant. His career path threaded through some of Westcoast's brightest and toughest ideas. He headed Pacific Northern Gas in its founding formative years. He worked on the Foothills pipeline project. And he was the champion of the Vancouver Island line. When the 1964 project died, he went on to other things, but the idea never left his head.

In 1968, Charlie Hetherington had his crack at a project called The Western Loop. He proposed a split line from Williams Lake to Powell River and Howe Sound. Hetherington wanted a second line into Vancouver, to provide security of supply and to lower the cost of the Island crossing. Politics prevailed when BC Hydro, which distributed gas in Vancouver, refused to support it.

Two years later, Rutherford was again in the picture backing a joint Pacific Northern Gas and Westcoast Transmission project from Williams Lake to Campbell River, believing economics had improved. Competing proposals forced the government to call hearings, which had just concluded when the NDP won the 1972 election. No decision on the application was ever rendered.

In 1979, Westcoast had advanced a proposal by Ron Rutherford for a $750 million fertilizer plant project at Powell River in conjunction with a pipeline crossing to the Island. The project was to be developed by a partnership of Westcoast, Chieftain Developments, Union Oil and B.C. Resources. It would have provided an "economic linchpin" for a pipeline to Vancouver Island because the fertilizer project would have generated half the cash needed for the crossing. The province, angered by federal energy policy, which it connected to Westcoast because of the Petro-Canada relationship, never granted a formal review of the fertilizer plant, effectively killing it and the pipeline crossing at the same time.

Next, in 1981, British Columbia premier Bill Bennett mandated BC Hydro to build a line without competitive hearings. Influenced by his anger over the National Energy Program, Bennett aimed to cut the federal government, including the National Energy Board and Petro-Canada–dominated Westcoast, out of the picture.

In 1988, a new B.C. premier, the flamboyant and eccentric Bill Vander Zalm, asked his energy minister, Jack Davis, to see about Westcoast reviving its northern route and ordered Davis to exclude BC Hydro from any and all consideration. By this point, Vander Zalm knew, the federal government was ready to help finance the project. In 1981, Ottawa had promised $500 million for each of two projects, one to move gas east of Montreal into the heartland of Quebec, and one to move west into Vancouver Island. The feds were still willing to come across with something.

Davis was a steady politician with a strong background in energy and resource issues plus federal and provincial cabinet experience. If the Business Council on National Issues had a textbook politician, it would have been Jack Davis. The problem that Davis faced, however, was that John Anderson was heartily sick of the Island pipeline and its politics. Westcoast had wasted money, energy and years of effort on an idea with sub-marginal economics. Instead of calling Anderson, Davis invited Rutherford, who was working at the time for Chieftain Developments, an Edmonton-based oil company controlled and run by Stan Milner, to Victoria for a meeting. The two were great personal friends who had played on the same university basketball team, and Davis knew that only Rutherford's enthusiasm for the pipeline exceeded his own.

"Why isn't BC Hydro building its line?" Davis asked Rutherford.

"Because it can't be done," the engineer replied.

"What do you think we should do?" Davis responded.

A few days later, Rutherford turned up in Arthur Willms's office to pitch his solution. Willms didn't want to know about it. "John Anderson will kill us if he even knows we're talking about this," he said.

Nevertheless, Rutherford unrolled his maps that showed a route starting from the Westcoast main line at Coquitlam, east of Vancouver in the Fraser Valley. It used Texada Island to make crossing the strait more practical and landed on Vancouver Island at Comox. "Just get the engineers to do a little work on it," Rutherford begged.

Secretly, Willms had engineers Jack Kavanagh and Harvey Permack work the numbers and the engineering concepts. With their endorsement, Rutherford went to a B.C. cabinet retreat at Nanaimo where he was promised that Westcoast would be given a ticket to build the line without hearings. Even John Anderson was reluctantly impressed, though he did not live to see the project launched. In the outcome, competing proposals emerged, and there was a public hearing that delayed the project for a year, adding $90 million to its cost.

By the time the project went forward, Alberta Energy Company had acquired Chieftain Developments, and was Westcoast's partner in a project company called Pacific Coast Energy. The federal government came through with $100 million and the British Columbia government backstopped certain costs related to the burden of getting the local distribution system up to full capacity.

Ottawa's contribution came as part of the Progressive Conservative government's 1988 election campaign. Before endorsing the deal, Phelps felt uneasy because the project's economics were too reliant on government largesse. He allowed himself, however, to be swayed by Jack Davis's enthusiasm and the Tories' "hands-filled-with-money" overtures. On the whole, he regarded Davis's judgement as the more reliable of the two.

Almost at the last moment, in one of those classic B.C. political moments, there was an election and the brash, impetuous union boss Glen Clark came to power as premier. He threatened to tear up the provincial financing commitments, but, as became the case so often in his chequered political career, he balked and the project moved forward.

There were some heart-stopping moments. Jack Kavanagh was supervising the final marine survey work from a hotel in Power River when, one afternoon when he was at his desk, there was a knock on his door. He answered it to find a brown envelope containing a BC Hydro study conducted for power line routes that indicated it was impossible to build an undersea facility south of Texada Island, where Westcoast proposed to lay its line. There was a dangerous potential for high-speed, destructive turbidity flows triggered by massive underwater coastal land slumps.

The secret delivery caused much consternation, but when Kavanagh had the route re-inspected he found no sand deposits from previous slumps, only slow-growing sponge beds that indicated life on the sea floor was tranquil and undisturbed. He faced a greater problem punching the line over the Vancouver watershed north of Coquitlam. There, rigorous restrictions were imposed on construction practices, including medicals for all the workers, to protect the major water source for the Lower Mainland.

The project encountered the classic challenge peculiar only to development in the Lower Mainland of British Columbia, whose citizens want the fruits of development without the burdens. Coquitlam mayor Louis Sekora played on this sentiment—raising a storm over the egregious notion that a pipeline for far-away Vancouver Island would pass through the nearby watershed. Incomprehensibly, logging operations being carried on in the same watershed didn't bother the folks who couldn't countenance the pipeline.

The furor Mayor Sekora created was aired at the regulatory hearing and the company successfully made the case that the pipeline could be constructed safely and without damaging the environment, which proved to be true.

Another demanding element was laying the line underwater. Westcoast engaged the dynamically positioned, vertical reel ship *Apache* to do the job. The *Apache*'s reels could hold 24 kilometres of pipe at a time, so Kavanagh had 83 strings fabricated at a work yard near the Vancouver airport. The *Apache* reeled them up one at a time, took them out to the right-of-way and unreeled them into the water. The whole operation took only 30 days, though the maximum water depth exceeded 1400 feet and at the time was the third-deepest gas pipeline crossing in the world.

As built, the pipeline system comprises 375 miles of 10-inch, high-pressure pipe from Coquitlam along the coast to Powell River, across the water and Texada Island to a landing on Vancouver Island north of Comox, where it splits north to Campbell River and south to Victoria. The project was badly overspent, climbing from its original budget of $260 million to $370 million, mostly attributable to delays during the public hearings. Westcoast had lost the bid to develop the local gas distribution system to Centra Gas, but some of its executives, notably chief financial officer Graham Wilson, didn't think that was such a bad idea because the economics of the retail operation were going to be pretty skinny.

Alberta Energy Company eventually sold its 50 percent interest in the pipeline to Westcoast. Later the company acquired the distribution system by taking over Centra Gas B.C. Two men who had brought the pipeline into being after thirty years of political intrigue and engineering challenge did not live to see the outcome. John Anderson and Jack Davis both died of cancer, Anderson in 1988 and Davis in 1990.

IT TOOK TIME TO consolidate the ICG and Union Gas acquisitions and integrate them into a business structure. Corporate functions were rejigged to reflect new roles and responsibilities. There was a significant internal reorganizing at Union, as well as expansion of pipeline capacity in British Columbia. There were new projects on the table in the spring of 1995; the momentum could be felt throughout the company. No one saw the cloud, no bigger than a finger, on the horizon. Across the continent, deregulation of natural gas transportation and marketing was accelerating. The storm hit Westcoast when it was least expected.

The story of the never-ending battle (range war would be a more apt description) between pipelines and producers is based on a simple equation. Each side regards the other as a necessary evil. Yet they desperately need one another. Without the pipeline, the producers would be stuck with capped wells; without the producers, the pipeline would be stuck with rusting, empty steel tunnels—in Westcoast's case, 8000 miles of them.

Every season it was the same. Gas producers and distributors, who knew very well what the company did, spent most of the year preparing for National Energy Board hearings on tolls and construction permits. In front of the tribunal, they attacked Westcoast with great zest and determination. Like all pipeline companies, Westcoast maintained a full-time staff of experts who annually prepared their case against the company's customers. It was, any fair-minded witness would conclude, a situation similar to that of the King of Siam, who, as recorded in *Anna and the King of Siam,* examined the ways of his country and declared them to be "a Puzzlement."

That changed, recalls Mike Phelps, on one very long day in May of 1995 very much like the day in 1954 when Frank McMahon, sitting in the barber's chair in New York's Park Plaza Hotel, learned that the Federal Power Commission had rejected his bid to build a Canadian gas export line into the U.S. Pacific Northwest and decided to try again.

In 1995, the company was riding a gas demand boom that impelled a $650 million expansion of its Grizzly Valley processing facility at Fort St. John and a related half-billion dollars' worth of pipeline looping and new gathering lines. Gas producers in northeast B.C., hurting from the lowest gas prices in years, thought otherwise. BC Gas, the province's distribution utility, also objected for competitive reasons, including protecting its own tolls. It seemed more logical to the producers to tighten gas supply in the expectation prices would rise. And, because they were netting only 80 cents per thousand cubic feet of gas they sold, it was all but impossible to pay the increase in pipeline tolls that expansion would bring.

When Westcoast had applied to the National Energy Board for a certificate to build the new capacity, the producers and BC Gas objected and intervened to force a hearing—on the grounds that the NEB lacked jurisdiction to regulate facilities that would be built inside British Columbia. The board agreed with them, and threw the project back at Westcoast. The company launched an appeal but also realized they had just been advised in no uncertain terms that their infrastructure in B.C. was now mature—that the producers were finding other ways to process their gas and didn't require more Westcoast capacity. By February 1996, when the Federal Court ruled that, indeed, the NEB had jurisdiction, the expansion had been shelved and Westcoast ate $45 million in development costs.

In the meantime, Westcoast looked at the way it did business and determined that it must restructure internally and retool. It also looked at its future, in terms of opportunities and investments to grow the company, "which turned out to be the best thing we've ever done," Phelps said after the dust settled, though neither he nor the stock market felt that way at the time.

And the change in the way the company did business also transformed its internal culture and the relationship between employees and the sheltering employer who had once been dubbed "Mother Westcoast."

Irv Koop, president of the pipeline and field services division at the time, recalled "We lost a billion dollars of rate base growth that day with the monkey wrench into our expansion plans. But we got the message. Our expansion was coupled with very depressed gas prices at the export point and linked to the perception of arbitrarily increasing tolls for service on our system. Customers wanted to leave our system and find cheaper alternatives."

Westcoast restructured the way it did business through a process called "We Change to Win," which amounted to cutting costs by 30 percent and laying off 25 percent of the company's workforce. The farther-reaching change was retooling how the company did business with producers and shippers. In a regulated world, the producers and the pipelines are enemies, just as the Prairie farmer and the railroads once were. Now, Westcoast invited its former adversaries, who had become its crucial customers, to help change the way business was done.

Here was how Irv Koop explained it. "Regulation is the surrogate for competition. The regulator was our customer. What the producers wanted instead was competition. And that was good for us, because, if you've got competition, you don't need heavy-handed regulation any more. When you don't have a monopoly or a franchise area any more, you need less of the traditional type of regulation."

The fundamental change, said Mike Phelps, was that, "we started to negotiate pipeline service tolls with customers rather than having them set by the NEB. We also negotiated processing fees. There is a floor to support the tolls, but the return the pipeline makes is dependent upon our efficiency, our reliability and the price of gas. We tied our fortune to the producers. We gave up the guaranteed rate of return except for a floor, which is essentially half the allowed usual return. The benefit for us was the flexibility to react to competitors, and to lead the market in a competitive world we needed the ability to react quickly. We needed that flexibility."

The new process was so successful that it became ongoing. Westcoast spent eighteen months negotiating a dramatically new way of doing business with its customers: the producers, the shippers and the marketers. It resulted in negotiated toll settlements for the pipeline and opened a new era of what became known as light-handed regulation.

In effect, Westcoast Energy was taking on a lot more risk for the chance to get a bigger reward. It gave up half its guaranteed earnings capacity in field services—about $26 million a year—and worked that into lower tolls, about 5 percentage points less than what it had been entitled to under the old model. The toll was then linked to the gas price received by producers so that Westcoast had the ability, depending on gas price, to make anywhere from zero to $90 million a year in earnings. It was a complicated formula but a simple idea: from now on producers and pipeliners would share good times and bad times together.

It was also a model that declared if Westcoast didn't provide its customers with high industry norm or world-class reliability standards, it would pay them money back. Each year the reliability figures were to be negotiated, and if Westcoast didn't meet them, it would pay refunds. Under the old model, a pipeline might not give good service or reliability. Not only did the producer not get the service, but a shortfall in revenue from that reduced service went into next year's toll. So the customer not only couldn't produce all his gas in a year that the pipeline performed poorly, but also had to pay higher tolls the following year.

Westcoast also took all the underutilization risk of $1.5 billion worth of gathering and processing assets. It was the first time anywhere in the world that a utility had taken on the risk of stranded costs. That is, if people leave the system, normally the utility still gets paid because the rates go up for service to other people. What light-handed regulation instituted was that if the pipeline was empty, Westcoast had no recourse but to have its $1.5 billion of assets at risk. What the pipeline got in return, which was so valuable, was the flexibility to do one-off commercial arrangements with its customers, allowing it to tailor services to what the customer wanted. Those deals could be done quickly because time was of the essence when the company was up against a keen competitor. Under the old model, that kind of deal-making required hearings and delays. The producer and pipeline had to wait nine months, twelve months or even fifteen months to go to a hearing, and then face interventions from other producers and even the competitor they might be trying to beat.

Now Westcoast simply reported its deals to NEB, stating the range of tolls and how they could be adjusted. There was still an element of regulation that gave the pipeline the ability to deal with the provincially regulated gathering and processing competition on a level playing field.

"We have been able to say, for example, that from basically gathering gas, processing it and transporting it there are about a hundred different products and services that we could offer, some existing and some new," explained Irv Koop. "We say to these customers, 'Here's what we can do for you. You tell us what's important to you and we'll bundle those services for you.' It's not dissimilar from where the airlines go with seat sales, or what the telecommunications or car rental businesses have done. Basically, what you do is price the product in relation to your inventory. If you book it two months in advance, the seat is $150; if you book it the day before the plane leaves, it's $980, and if you come when the gate closes, they'll give the seat away. For example, a shipper may have a customer who wants 10 million cubic feet of gas each day in an eight-hour time frame [for commercial air conditioning]. We're already moving the 10 million in a twenty-four-hour period for the shipper, but he's willing to pay a premium if we move it in that eight-hour time frame. That's revenue management.

"To add value to the assets is what we call strategic account management. The strategic account management is one in which we see the opportunity to create value for that account and ourselves over the long term. They become the premium accounts," Koop continued. "It's a whole new way of doing business, and we've taken on a great deal of risk with it, but it began to produce record financial performance in the pipeline and field services divisions right away. One reason was that gas prices were very strong in 1997 and 1998, and since some of the income was linked to gas prices, we had an ability to get some pretty good returns and to pay for the pre-investment of dollars that went into the retooling."

NO MATTER HOW one added it up, the NEB decision in May 1995 to reject Westcoast's expansion plan for northeast British Columbia prompted the company to look elsewhere for its next opportunities.

It was very much like being thrown out of home, and the trauma produced a new agenda: light-handed regulation, an internal retooling called "We Change to Win" that would transform Westcoast's corporate culture, and the decision to look over the mountains eastward and west around the Pacific Rim for places to invest and grow. Westcoast would seek a North American footprint and a global toehold.

15

Transforming Corporate Culture

Westcoast changed dramatically in style, geographical scope, integration and diversification. There is one constant—energy in all its forms and markets.

ED PHILLIPS

Beyond its inanimate balance sheet and dry operational detail, a company is a creative act—a genetic tapestry of human invention, onto which new threads are continuously spliced and woven. The resultant organism contains, in its DNA, an encoded record of the corporation's past, the architecture of its present culture and style, and the trajectory of its destiny. This completeness of past, present and future differentiates the company from its competitors, animates its quest for profit, and commands respect, loyalty and affection in the people who own shares in it and work for it.

Companies begin as a legal act recorded in documents that soon gather dust and are forgotten. Companies come to life in transactions tangible and intangible, commercial, social, financial, scientific, intellectual and physical. Companies express their identity in their geography, physical structures, internal architecture and external declamations. But corporations also live through owners, employees, customers, suppliers, neighbours, regulators and legislators. They can claim existence through all the careers they affect. At their core, companies are the sum of their relationships, experiences, defining moments, benchmarks and watersheds. They are buffeted by external forces and occasionally torn by internal conflict. Some relationships endure and some relationships end, but all relationships change. No company is immutable. Companies change, and that change is deeply personal for those whose lives they touch.

In 1996, responding to irresistible external forces, and facing choices upon which its competitive survival depended, Westcoast Energy's corporate culture experienced the most traumatic shift in its half-century history. It was called We Change to Win—a twist of the phrase converts that to Winds of Change, and truly Westcoast was caught on the open sea in a monumental tempest, a Force 10 gale. Rather than run for shelter, Westcoast turned, bow first, into the teeth of the storm. To understand how dramatically We Change to Win altered Westcoast Energy's culture, one must grasp something of the way things were before the deluge.

PIPELINE COMPANIES ARE bounded by their geography, and before 1996 the map of Westcoast Energy's world was the skeletal structure of British Columbia. Its pipelines ran from a hub of processing plants at Fort St. John, diagonally across the mountains of the Cordillera, and along their interior valleys and rivers to Vancouver, the Lower Mainland and the American border. The web of steel had a northern extension to a processing hub at Fort Nelson and beyond to new gas fields at Liard and Pointed Mountain in the Northwest and Yukon Territories. A second web wove westward to the sea at Prince Rupert. A finger poked across the Strait of Georgia to Vancouver Island. There were outposts of oil and gas lines across the Alberta border. But the map had an underlying message: Westcoast lived in a sheltered world that it dominated. In Vancouver, Westcoast was the leading corporate blueblood. At the northern end of its physical world, Fort Nelson still was a company town and Fort St. John, a Westcoast dependency.

At the outset of its operations in the Peace River country, the company provided the best wages in northern British Columbia. It built housing for its employees, rented at low rates and provided a generous northern living allowance. Employees in the field might be members of the pipeline industry's ubiquitous trade unions, but they also felt part of a family, in the best tradition of the Canadian resource company town with its company-store umbilical cord of supports.

Retired electrician John Anderson recalls that when he moved to Fort St. John in September 1957 to start work at Station One, "there weren't accommodations so my wife, myself and five children lived in a single motel room for six weeks while they built our house. It was ready about the middle of October and the day before we moved in there was a violent snowstorm. I came from the dry, warm Kelowna area, and when I went to the house that first morning, there was snow drifted up to the door. There was no telephone. If my wife wanted to call anybody, she had to go over to the company dispatch office and use the phone there."

Anderson recalls that Westcoast families were isolated from friends and relations in the south and had to stretch family budgets to pay for the higher cost of everything from milk to winter clothes. The company initiated a northern living allowance to help with the cost of living, and co-workers became surrogate extended families, but the frontier provided its own compensations: spectacular hunting and fishing, family campouts and picnics, hockey and curling.

It was a cradle-to-grave environment. Most employees were effectively guaranteed their jobs for life. And if the people had certain expectations of the company in terms of their career, the company expected their loyalty in return. If there was an incident, a leak in a major pipeline, for instance, the company knew its people would work day and night to fix it. Mother Westcoast ruled the roost.

Frank Parker joined the company in August 1965 and stayed for thirty years. When he retired, he was a corrosion technician at the Fort Nelson plant. "In the early years, they let you do your job. It was miserable, cold and dangerous, but everybody went out and did what had to be done to keep the plant running. I always figured that if the company didn't like how I ran it, then they'd run me off, but in the meantime I was the boss." It was a command-and-control environment, but it was also Western Canada's best technical training school, fostering self-reliance and job pride.

The Vancouver headquarters was a different world. In its history, the company had more than its share of corner office politics, but the often tense atmosphere was leavened because executives frequently took off to Fort St. John or along the line. Although Ed Phillips did it best, all the CEOs and senior executives made an effort to get out to events such as local bonspiels and Christmas parties that highlighted company social life. The self-reliance, the pride in wearing the company logo, the sense of doing something important for the community provided the company's cellular structure. The employees in the field had some appreciation of the internal struggles in Vancouver, and the employees in the city absorbed the free-spirited loyalty of their counterparts along the pipeline.

Time and again, executives and employees who lived through the Mother Westcoast years had a simple description of the corporate culture before 1996: it was family. Frank and George McMahon founded the firm as, literally, a family business. The Peace River country experience endowed the culture with a unique sense of kinship nurtured by Ed Phillips's avuncular style. But the flip side of the cozy workplace was an underachieving company—and an organization that was eventually sideswiped by larger economic forces.

WHEN THE $1 BILLION pipeline expansion in northeast British Columbia collapsed in 1996, Westcoast Energy faced a major overhaul of its pipeline and gas processing business. "We changed two things," recalled Irv Koop. "Our traditional enemies became our customers. And we rewrote the employment contract."

Koop recounts that Westcoast's executives bit the bullet. "We did the right thing. We restructured the company. We took three hundred people out of the organization; that was 25 percent of our work force. It was quite a cultural shock because it was the first time in forty-odd years that the company had terminated people without grievances. The reason we were able to do it is that the crisis was big enough. Employees could watch people building plants next to ours and taking our business away."

"It was the end of the family company," recalled Dave Unruh. "It was very traumatic, a watershed event for the employees, for the senior executive team and for the CEO. The company had to do it. In the modern business age, in a company with a worldwide base, it is very hard to preserve a family-type company culture."

"We wiped out some departments and, in others, none of the existing people stayed. It was a 25 percent job cut overall; at the unionized level, the number was smaller than 25 percent, and in other departments 50 to 75 percent of the people left," observed human resources vice-president Bohdan Bodnar. "It was a cultural shift to move away from the regulated mindset and to move closer to the customer mentally and sometimes physically. It focussed the attention of people on the needs of the customers because that was how we made our money now."

Westcoast would be staffed for the valleys of its business cycle, not the peaks. Many jobs were contracted out. Even when tasks were kept in-house, administrative units had a service agreement with the core company—literally paper contracts with "fees for service."

"We Change to Win led to the cultural issues that we are still going through," Irv Koop admitted. "It takes time to move from a utility mindset to realizing that customers are paying the freight and you've got to be responsive to their needs and wishes."

The main reason the trauma was accepted, however reluctantly, by the employees, including the long-serving, departing ones, was that it drew on the credibility of Art Willms. He had built his career on his integrity with external stakeholders as the company's chief witness for twenty-five years at the National Energy Board. He had been the company's most successful career mentor. The people who worked for him were damn proud of him; he was their role model of how a good Westcoaster did his or her job honourably. So when Art Willms, as the executive team leader, said, "this has to happen," his edict was received with surprising equanimity.

The employees who kept their jobs were troubled, suffering from survivor's syndrome. It took people a long time to heal. The depth of the anger and pain became apparent in the Great Turkey Caper, an E-mail war that broke out when the company decided it would no longer distribute Christmas turkeys to all employees. People felt it was the last straw for management to trim the work force 25 percent, and now it was stealing the Christmas turkey.

Paradoxically, however, the E-mail revolt seemed to reaffirm the familial values now so deeply ingrained that they were passed on through new iterations of the company. Art Willms insisted that the quality of family remained. "Other outside observers believed that Westcoast Energy still had a culture that's very familial. The evidence? When there was a crisis, the people still banded together, really quickly and really easily, and rose to the occasion."

"Westcoast," said Bohdan Bodnar, "was looking for people who could create an 80 percent solution to a problem on their own initiative and with their own knowledge and skill and who are normally right. It was no longer the kind of environment where you can afford to make a mistake. The regulated culture was more forgiving. Now you had to take risks. People describe the old culture as 'Ready... aim... aim.' The new culture was, at its best, 'Ready... fire... aim.'"

Dr. John Sehmer, who has spent years caring for the health of Westcoast employees, noted that the company was slow to adapt. And he sounds a warning: "The contemporary wisdom is that you no longer worship people's duration with the company, you worship their productivity. That's a terrifying concept and a fashion that I think is going to disappear. It is basic human nature that you need to have a sense of rootedness. What most of the management gurus have been writing recently is that everything and everyone has to be value-added, that there ought to be massive out-sourcing, that people's duration with the company is far less important than their productivity, and that you should staff for the valleys, not the peaks. Westcoast was forced to move in that direction and as a result people don't feel the sense of belonging they once did. It's a tough market, with shareholder returns becoming mission statements. It was a difficult balance for Westcoast because they're dealing with a non-renewable resource and have to weigh immediate earnings with long-term supply and demand."

We Change to Win all but eradicated Westcoast's traditional sense of security. While the employees were absorbing the change, Westcoast emerged as a company with a North American footprint built around four geographic centres: British Columbia, Ontario and the U.S. Midwest, Atlantic Canada and the U.S. eastern seaboard, and Mexico. Union Gas was being run as a virtually separate fiefdom. The company had grown so big and so scattered that it seemed unimaginable that the family culture would ever be recovered. Not all of the new generation of employees even wanted it.

What remained, observed Mike Phelps, was a culture of deal-makers and engineers that had been evolving since the acquisitions of Inter-City Gas and Union Gas. The original core culture was an engineering culture. And in the executive suites Westcoast had the deal-makers, the people with a strong appetite for the growth of the organization. "As companies grow through acquisition there are cultural irritants," he said. "When we lost the expansion decision, we had to negotiate something radically new with producers and get our costs down. The only way to get costs down in response to the crisis was to go at it extraordinarily vigorously. After the We Change to Win program, we needed to pull things together and rebuild relationships. And that is a long, continuous process."

While Dr. Schmer observed, "When you had a job at Westcoast, you had it for life, and that day is gone," Art Willms believed, "We are still an extended family, even if we don't still get together for dinner every Sunday."

WEDNESDAY, JANUARY 27, 1999, was a quiet afternoon on the streets of Taylor, the British Columbia town that Westcoast Transmission and its partners put on the map in 1957 with construction of the energy cornerstone of its Peace River country development plan. For forty-two years, the giant McMahon gas plant, built by Pacific Petroleums at Taylor, processed the continuous torrent of natural gas that flowed into the Westcoast trunk line travelling south. Associated plants refined BTU-rich petroleum condensate into aviation gas, and other facilities extracted sulphur, propane and butane. Taylor was the hub of the economic miracle that Westcoast created on the province's northeast frontier. The angular towers of refineries and gas processing vessels that make up the skyline spewed clouds of steam into the frigid sky. The lunch hour had just begun, but the structures threw lengthening afternoon shadows across the snow: night comes early in midwinter above the fiftieth parallel of latitude.

At the Solex Energy gas liquids plant, a service crew was calibrating a butane meter when a faulty hose coupler broke loose, loosening a high-pressure flow of gas. The errant line whipped and snaked out of control until the inevitable spark touched off a massive blast. With sudden brute force, three quick thundering roars cracked the air and the ground shook. First a pillar of black smoke belched out, and then a fireball erupted 200 feet into the sky and flared out into a wall more than 300 feet across. The surprisingly fragile $100 million gossamer of pipes and cooling towers that made up the Solex plant simply disappeared in the continuous roar of burning methane that poured into the atmosphere from the fatal wounds the plant had suffered. Within minutes, the skirl of sirens and the pulsing blue and red lights of emergency vehicles had torn the afternoon somnolence into shreds. Loud-hailers paged the town's one thousand residents to a hasty evacuation.

Al Laundry was the manager of Westcoast's Taylor complex, which includes the McMahon plant, an electric power co-generator and Station One, the compressor system at the northern head of the Westcoast main line. He was sitting down to lunch across the parking lot from the Solex plant when the emergency siren howled. Laundry's heart leaped into his throat; then he relaxed—Wednesday noon was the weekly scheduled time to test the plant's alarm system. Then he and his men heard a second siren, this one from an emergency vehicle. As Laundry raced for the plant's emergency base station he felt the punch of another explosion. He listened grimly to the shouts and commands coming from the Solex plant on the emergency radio system as the biggest explosion in the sequence rocked the floor. "By this time, the flames were already visible over the buildings. This was extremely dangerous, all these explosions. You had to yell to make yourself heard; the windows rattled and the ceiling tiles shook and glass showered in on us," he recalls.

Then came the worst moment, as a huge wall of flame shot up and radio communications suddenly went dead. "I knew from talking to our men in emergency response that they'd gone into the Solex property and were fighting the fire," he vividly remembers. "I've gone through a few of these things and knew that with that amount of radiant heat, our men were in extreme danger. I couldn't get any communication from Solex's people or the guys with the fire truck. I announced on the radio that we would take over the emergency response."

By now a mist, possibly of gas and liquids, was drifting off the Solex lot towards the Westcoast plant. "These kinds of situations tend to get progressively worse and I thought that I might have lost my whole emergency response team and I wasn't going to lose anyone else," he recalls. Laundry ordered an emergency shutdown of his own plant, and a crew went around battening down the hatches.

As quickly as they could, the Westcoast team evacuated their own buildings. As the cold seeped into the unprotected facility over the next few hours, it suffered millions in damage as steam lines froze and cracked, and molten sulphur turned to stone. There was no option, with a half-billion-dollar facility to protect. Westcoast's own plant was not the main concern of its people. There was a downed crew to rescue and the hometown to protect. When their own plant was locked down, they went to the Taylor community centre and took command of the job of containing the fire and protecting other natural gas facilities. At the edge of town, RCMP officers counted the fleeing townsfolk, to make sure everyone was out, and sent the temporary refugees down the road to family or friends in nearby Fort St. John.

Unbelievably, no one had been killed, but the crew was down with a variety of burns and blast injuries. In the heroics of the rescue operation, several emergency personnel were felled, including men from Westcoast who had been blown right off their truck and had their jackets burned off by one of the secondary explosions. In the next ten hours, until the evacuation order was lifted, as crews worked to contain the crisis, Westcoast's well-trained, well-equipped and field-seasoned personnel played a stellar role. They attacked the blaze in a series of small, coordinated blows until eventually all the fires were contained, and all possible sources of gas shut down. Although small fires burned in spots for two more days, the emergency ended and the alert stood down at 9 p.m., an astonishingly rapid response to a major catastrophe. It took a week of round-the-clock work to assess the damage, restore a semblance of order to the affected plants and start repairs. It took a full nine months to wipe out the last trace of damage to the Westcoast plant.

Steve Thorlakson, the mayor of Fort St. John, was one of the local officials called to the scene to help. Seeing Westcoast at the centre of the action was reassuring amid the disaster. Once again, the company was playing the role of Peace River country cornerstone. Thorlakson has been in Fort St. John for more than twenty years and the Westcoast employees he first met were members of a "pretty paternalistic organization. The kids who got jobs in the summer were kids of the local managers or senior employees. There was very much of a family feel because it was a relatively modest-sized company compared to what it has become."

Thorlakson heard tales of the era, twenty years before he arrived, when Fort St. John and Fort Nelson were Westcoast Transmission and Pacific Petroleums' company towns. He cites the special relationship between Westcoast and Fort Nelson as a benchmark. "They didn't have a sound community and well-financed facilities, and they had a horrendous 40 and 50 percent employee turnover because there weren't facilities to keep families there," he recalls. "So there developed a relationship between Westcoast and the community where Westcoast invited itself to be taxed to help pay for the facilities." In 1980, Fort St. John went legally bankrupt, though no one petitioned the city into receivership. "We looked at Fort Nelson and Westcoast and said, 'That should work here, except that this town with ten thousand people has more to offer.'"

Unfortunately, before he could negotiate a proposal with Art Willms and Ron Maas, the NDP's provincial finance minister, Glen Clark, slapped a $6 million tax on compressor fuel, which earned the nickname "the Westcoast tax" because it was aimed at one company. "Westcoast, which was already the largest single property taxpayer in British Columbia," Thorlakson said ruefully, "was somewhat less receptive to our plight after just having been kicked in the groin by Glen Clark; 80 percent of their tax burden goes to Victoria."

Fort St. John survived its crisis with the help of good advice and hand-holding from its biggest corporate citizen. Westcoast also provided tangible help with its investment in and promotion of Alliance Pipeline, which originates its 3686-km big-inch pipeline to Chicago at Fort St. John. British Columbia's oil and gas commissioner, Rob McManus, says Alliance requires 600 million cubic feet per day of new natural gas production out of B.C., representing a 30 percent increase over 1998. By 2008, he expects the Peace River country's gas production to increase by 50 percent over 1998. Completing the Alliance pipeline and drilling enough wells to fill it is pumping $2 billion a year into the Peace River country economy—a fifth of the petroleum industry's typical annual spending in Alberta. Once again, Westcoast Energy's leadership as well as its direct investment is the catalyst and cornerstone of the Peace River country's next stage of development.

Mayor Steve Thorlakson summed it up: "The Peace River country's relationship with Westcoast Energy has changed over forty years. We're no longer a wooden-sidewalk, brawling, mud-and-blood-in-the-street, boom-and-bust town. The energy development and production industry has taken a lot of the peaks and valleys out of its system. But we still have a tremendously strong relationship with Westcoast Energy."

16

The North American Footprint

Yes, I'm a stranger to Nova Scotia,
This land so pretty to see,
Yet I'll have to pull so hard,
To take my heart back home with me.

GLENN HART

The discovery in the 1970s of natural gas on the Scotian Shelf, in the graveyard of the North Atlantic where hurricanes and pirates once drove ships to their deaths, transformed the North American natural gas map and opened the door to a new economic era for Maritime Canada.

The first era, which lasted two hundred years, saw the region's shipyards, fleets, fishermen, lumberjacks and miners make the region wealthy and proud. There were, of course, other resources: thick forests that fed sawmills and pulp mills on both sides of the Atlantic, black soil and neat orchards that yielded crops in abundance and coal drawn from deep beneath the earth.

The second era was a time of testing and trial. The region fell behind the rest of the nation and became a political dependency as entire industries shrank or dried up and the fisheries faced oblivion. Only an unquenchable optimism and the indomitable qualities of the human spirit preserved communities and maintained hope and confidence. No Canadians love their region more deeply or more loyally than Maritimers love the Maritimes. Everyone else in the world is "from away." But in spite of the words of its great musical interlocutor Rita MacNeil, "Banish thoughts of leaving, it's home I'll be," young Maritime fathers went to work on the drilling rigs and in the oil sands plants of Alberta, becoming the biggest blue-collar workforce in oil sands capital Fort McMurray, Alberta.

When oil and gas was discovered on a pearl necklace of basins from Newfoundland to Maine, there was an almost imperceptible freshening of the wind. The regional economy had been quietly gathering the critical mass of a renaissance since the mid-1960s. Alberta had the American oil companies to kick-start its petroleum era in 1945. The Arctic north had southern oil, mineral and diamond investors. British Columbia had the Japanese and Koreans. But Maritimers had only themselves.

The renaissance started slowly and gathered strength invisibly between the late 1960s and early 1980s. It was created and led by Maritimers like Roy Jodrey, Frank Sobey, the Irvings, the McCains, Frank McKenna and Robert Stanfield. In Prince Edward Island, the turning tide produced the causeway. In Nova Scotia, a new generation of businesses arose, including international food-processing industries and investment machines like IMP Group, Scotia Investments and Maritime Life Insurance. In New Brunswick, it was the capital might of homegrown K.C. Irving Limited that made the most of New Brunswick's potent shipbuilding, oil refining, trucking and bus and rail services. A new generation of entrepreneurs carved a New Brunswick niche as Canada's preferred call centre to take advantage of a bilingual population and attractive economic development policies.

For two decades, the discovery of natural gas near Sable Island off Nova Scotia seemed to mock the Maritimes, dangling and withholding the prospect of a new source of wealth and a new sense of direction. The decision to bring the gas ashore, integrate it into the regional energy economy and export it to New England became a symbol of the renewal of the Maritime economy, a zephyr that encouraged a change of course. The Maritimes & Northeast Pipeline company was to become a cord of steel that bound Nova Scotia and New Brunswick together.

In 1995, the Sable Offshore Energy Project (SOEP), which owns the trillions of cubic feet of gas in six fields on the Scotian Shelf, east of Sable Island, decided that natural gas prices and market demand were good enough to produce the fields. SOEP had four owners: Mobil Oil, at 50 percent; Shell Canada, at 32 percent, and Imperial Oil and Nova Scotia Resources, at 9 percent apiece. In June 1996, the producers applied for regulatory approval to install six offshore drilling and production platforms, complete thirty wells at distances ranging from 160 to almost 320 kilometres off the coast, and construct more than 400 kilometres of gathering lines to bring ashore up to 500 million cubic feet per day of natural gas. The initial capital cost was $2 billion, and they planned to expand over a decade to 1 billion cubic feet per day.

The producers had to create a second consortium to build a $1 billion, 625-mile, 30-inch high-pressure pipeline to ship the gas through Nova Scotia and New Brunswick to Boston, Massachusetts. Mobil Oil and Houston-based Duke Energy recruited Westcoast Energy to build and operate the Canadian portion of the system.

These three companies formed Maritimes & Northeast Pipeline, which was owned 37.5 percent by Westcoast, 25 percent by Mobil and 37.5 percent by Duke Energy. Taken together, the pipeline and offshore facilities would be the biggest construction project in Atlantic Canada's history. In October 1996, Westcoast, acting as lead in Canada for the Maritimes & Northeast project, filed an application with the National Energy Board for permission to build the Canadian portion of the line.

Pipelines are mostly politics, and the most political project of them all was the Maritimes & Northeast Pipeline. Not since the great pipeline debate of 1957 brought down the Liberal government of Prime Minister Louis St. Laurent, and ended the careers of the politicians who managed Canada's postwar economic miracle, had there been such a bitter conflict over pipelines. The economic future of Canada's

Atlantic region was at stake, Quebec's nationalist politics had ensnared the project, a federal election was in the balance, and Westcoast was facing down the main rival, whose very name, TransCanada PipeLines, reflected its belief that it had a divine right to build the line. Trans Quebec and Maritime Pipeline (TQM), co-owned by Quebec distributor Gaz Métropolitain and TransCanada, gained National Energy Board permission late in the 1980s to extend TQM from Quebec City to the Maritime provinces to deliver Western Canadian gas. But it had never acted on the project, which would have been uneconomic without massive subsidies. When Mobil Oil, Duke Energy and Westcoast proposed the Maritimes & Northeast project, TQM discovered its piece of paper had passed its "best before" date.

The opportunity to build the Maritimes & Northeast line came at the nadir of Westcoast's experience in British Columbia. In May 1995, the National Energy Board gunned down the $1 billion expansion of Westcoast's system in B.C., ending what seemed the last opportunities for the company to show its stuff as a builder of gold medal pipelines in the toughest terrain on earth. The company had been planning the project for a year; losing the chance was a blow to the pride of its people and touched off an overhaul of the company that cost a quarter of the workforce their jobs. Westcoast needed a dramatic event to get back into pipeline expansion and ensure its long-term growth.

"It was a watershed," said David Unruh. "There were other companies, but Mike Phelps and Art Willms had credibility in the North American community, and Duke Energy knew them well."

"We had to step outside of ourselves to do it," said Unruh. It wasn't the technical, financial or market aspects of the project that presented the steepest challenge. "Maritime politics are very interesting, and we had to adjust for that. We also had to adjust for competition from Quebec; Hydro-Québec and Gaz Métropolitain wanted to use Sable gas to create a hub near Montreal. It became a full-time job for a number of us to manage this project through the corridors of politics and government."

TransCanada PipeLines, TQM and the Quebec, Nova Scotia and

New Brunswick governments raised a lot of objections to a project that would export most of Nova Scotia's offshore gas to the United States and not deliver a cubic foot to Quebec. At first the political establishment didn't take either Mobil, the big ugly American, or Westcoast, the sleepy little pipeline company from the far side of the Rockies, seriously.

Westcoast knew how to deal with the National Energy Board and just plugged along, getting ready for hearings in 1997. The Maritimes & Northeast Pipeline would pass regulatory muster—of that there was little doubt. The opponents would fight it out in the political arena: in the provincial capitals and in Ottawa.

There was a dispute with the Atlantic provinces over tolling—the transportation charges applied to the gas on the pipeline. Maritimes & Northeast proposed a postage stamp toll, which meant all customers paid the same amount regardless of how far the gas travelled to get to them. The Nova Scotia government, in total opposition to the New Brunswick view, wanted toll charges lower in Nova Scotia than in New Brunswick. Both wanted higher tolls for American exports, which would serve to reduce tolls for Canadian buyers. And, at the eleventh hour, the Nova Scotia Mi'kmaq First Nation sought an enriched deal for the right to cross over traditional tribal lands.

But the big threat came from TQM, backed by the Quebec government. As the Maritimes & Northeast application went to hearings in Canada, TQM mounted a counterproposal. Instead of the line turning south in New Brunswick, it would continue east to Quebec City to join the TQM line, which would be reversed. The American gas exports would travel to Montreal and be shipped south on TransCanada's Portland line through Vermont.

The proposal was entirely political. TransCanada had no shipping agreements with the producers, who wanted the Maritimes & Northeast line. Its route would be heavily subsidized by tolls on the TransCanada main line, and it would turn the American market on its ear by sending the gas on a 600-mile detour. And TransCanada had not bothered to file an application to build the line with the National Energy Board. In retrospect the TQM proposal was pure mischief-making.

previous page: Its odd design ordered by Frank McMahon to circumvent land use density regulations to be Vancouver's tallest building, Westcoast's headquarters on West Georgia Street remained the corporate nerve centre until the company was acquired by Duke Energy. WESTCOAST ENERGY

above: It took the reel ship *Apache* just thirty days to lay the Vancouver Island pipeline from Powell River to Comox, forty years after Frank McMahon put the project on Westcoast's "to do" list. WESTCOAST ENERGY

right: A jewel in the Union Gas crown, the Dawn, Ontario, natural gas storage facility allowed Union to dominate the gas market in southwestern Ontario, influence distribution across the province and play a pivotal role in shipments to and from the U.S. UNION GAS

right: A pipeline construction crew building the Maritimes & Northeast project in 1999 to deliver natural gas from the offshore Sable Island field to Nova Scotia, New Brunswick and Maine prepares to conduct an internal electronic inspection on a completed section of the line. WESTCOAST ENERGY

below: A welder completes a link on the 3687-kilometre Alliance Pipeline from Fort St. John to Chicago. The pipeline provides an important market link to the largest new gas discoveries in western Canada and the southern Northwest Territories. Westcoast's investment in Alliance strengthened its North American footprint. ALLIANCE PIPELINE

above: The offshore nitrogen reinjection facilities for the Cantarell oilfield in Mexico's Campeche Bay were completed in 2000 by a partnership including Westcoast Energy. An onshore plant that is part of the complex produces high-quality nitrogen, which is injected into the oil reservoir to increase and prolong production.
WESTCOAST ENERGY

left: Participating in the 4.4-million-hectare Muskwa-Kechika land use protection area in British Columbia's Northern Rockies "was one of the best day's work Westcoast ever did," says Michael Phelps, the CEO who made the decision. It is the world's largest big-game preserve outside of the Serengeti Plain in Tanzania.
STEVE SHORT

The 50-megawatt Wei Gang power plant in Shanghai, People's Republic of China, produces electricity using waste heat from steel blast furnaces. It was Westcoast's smallest foreign foray but gave the company a toehold in a vast market. WESTCOAST ENERGY

The Canadian Association of Petroleum Producers dismissed it as, "an anachronism wrapped in a myth detached from any market reality."

Westcoast decided not to promote its project through partisan political connections, but to present it to the NEB on its merits as a market-based enterprise that didn't require government subsidies. As Art Willms recalled, "We had a great project; then our competitors came to the table late and started proposing a line to Quebec. We suffered great consternation when Prime Minister Jean Chrétien mused on two occasions about how nice it would be to have an all-Canadian line—but he was just thinking out loud, and while it caused some grief, it had no effect on the competing parties."

Maritimes & Northeast took a lot of time—five years from the first piece of paper to the final weld—because of the intense politics. TransCanada PipeLines, Gaz Métropolitain, and Hydro-Québec wanted to pull the line up through Quebec, effectively creating two sources of gas for Quebec, one from west and one from east. At a bilateral meeting with Quebec premier Lucien Bouchard, Prime Minister Chrétien said they should take a look at the possibility.

"That had a superficial appeal—one national pipeline. The difficulty was that the Maritime market was always going to be too small to develop the gas, bring it ashore then build a pipeline through Nova Scotia and New Brunswick to Quebec," said Michael Phelps.

Maritimes & Northeast president Pat Langan hated the idea and the delay. When journalists asked him to comment, he dismissed the prime minister's musings as impractical and uneconomic. The implication was that the prime minister should keep his nose out. This repartee created a diplomatic crisis for the project. Phelps, who first read about Prime Minister Chrétien's musing that TransCanada might be right at an airport newspaper stand while on a refuelling stop, got onto the phone to the prime minister to do damage control, while Arthur Willms took Langan aside and suggested he keep out of the media.

The bottom line was that Sable Island gas development needed the New England anchor, just as forty years previously, to develop B.C. gas, McMahon had had to find an American anchor market. Phelps observed that the producers were not interested in satisfying Montreal when Montreal was already being serviced with western gas. But the political power of economic interests in Quebec is very strong, and the federal government, for very good reasons, wanted to see Quebec further tied into the national economy.

"We spent a year meeting everybody who counted for anything politically in Ottawa. They would quickly agree with the economic rationale underlying our proposal, but nobody would say no to the other side," Phelps said. "We needed to get Nova Scotia and New Brunswick on side. But they had different agendas. Nova Scotia came on side with us quickly, but New Brunswick is of two minds when it comes to Quebec. But we ultimately got New Brunswick's support. Mobil was the key. Mobil said, 'We support only the Westcoast-Duke proposal,' but the other guys said to Ottawa, 'Don't believe them, they will shift to whomever gets the certificate.' Eventually, it got sorted out."

To get regulatory approval, the Maritimes & Northeast Pipeline proponents were put through the longest hearing in the history of the so-called deregulation era, including environmental review and the traditional technical assessment of engineering, construction and tolls. But the defining moment that Arthur Willms remembers took place not in the front room of the regulatory process but in the back rooms of a region with differing political interests and nuances.

"Nova Scotia and New Brunswick were fighting over the rates we were going to charge," Willms recounted. "New Brunswick wanted a 'postage stamp' system where everyone in Canada pays the same to receive a cubic foot of gas. Nova Scotia wanted it based on mileage because they were closer to the fields and wanted to pay less. I was afraid that if the National Energy Board saw these guys fighting, they'd delay a decision." A delay would strengthen the hand of TransCanada as it tried to derail the Maritimes & Northeast project and replace it with a pipeline to Quebec.

Meanwhile, Willms was doing his regulatory minuet. The project faced a 1970s-style contested hearing and Art was the one guy left active in senior management who had credibility and experience and enormous wisdom about the business. "He's a lot of steel wrapped in a gentlemanly package," Phelps explained. "TransCanada's main job is to ship hydrocarbons out of western Canada to midwestern and eastern markets. Had it done the Quebec line, it would have been in conflict with producers from Alberta, so we were the natural choice."

Phelps and Willms met regularly with Nova Scotia premier John Savage and Premier Frank McKenna in New Brunswick. It was shuttle diplomacy. Willms tried to move them to a common position, but the two men refused to meet face to face. Willms found it impossible to line up the provinces' positions—to get the compromise and consensus the project needed. Finally he insisted, late one night in June 1997, that Premier Savage get on a company plane with the Westcoast team and fly from Halifax to Fredericton. The light aircraft bucked and twisted through a storm and landed at 11 p.m. McKenna was waiting in his office. The two premiers rolled up their sleeves and talked—until 2 a.m. And they cut a deal so good that when Premier John Savage resigned and the new premier, Russell McLellan, made noises about re-opening the deal, Art Willms successfully held the agreement intact through another round of shuttle diplomacy.

"The outcome was the most satisfying in my career," said Willms.

In December 1997, the National Energy Board granted Westcoast Energy and its partners, Mobil Oil and Duke Energy, approval to build the line from Goldboro, on the east coast of Nova Scotia, to markets in Nova Scotia, New Brunswick and the New England states. The permit represented a triumph of Westcoast Energy's political craft as much as it did the logical outcome of an evaluation of the national, economic and social merits of the case.

The victory belonged to Arthur Willms more than to any other individual. It was the culmination and product of Willms's career as a "terrific operator, solid financial pipeline thinker and fine leader of people, and of his credibility in the industry and reputation as the world's best witness before any regulator," said David Unruh.

On February 11, 1999, a crew near Fredericton, New Brunswick completed the first weld on the pipeline. It was, noted Westcoast Energy's in-house newspaper, *Energy Times,* exactly forty-one years, six months and three days since the final weld was completed on Westcoast Transmission's British Columbia pipeline. A little more than five months later, in August, the last length of line was lowered into its trench and buried. After six weeks of testing, the pipeline was cleaned and filled with gas in October. "Five years ago this was an idea on paper. We said we would have the pipeline up and running in November 1999. And we've delivered," Mike Phelps emphasized.

WESTCOAST ENERGY'S $429 million investment in the Alliance Pipeline project is as close to a passive investment as any it made during the tumultuous project years of the 1990s. British Columbia and Alberta gas producers drove the idea and initially owned Alliance. They continued to drive it because they wanted to transform the market and create eastbound transportation competition after years of frustrating dealings with TransCanada PipeLines. Westcoast joined the project after it was already well advanced. It was one of five pipeline companies that hold equity and it is the second-largest partner. And its investment of nearly one-half billion, though the single largest exposure it has taken since acquiring Union, was relatively straightforward to finance because of the strength of Alliance's long-term contracts.

That said, Westcoast has contributed more than money; it was also the Canadian participant with domestic natural gas construction and operating experience, and Westcoast assigned some of its talent pool and all of its expertise to the project. The company had much at stake in a project that is revolutionizing the marketing of Canadian gas and will play a role in bringing reserves out of the Arctic.

At the time construction was completed, the major owner was energy trust Fort Chicago Energy, and the partners included Westcoast, Canadian oil pipeline giant Enbridge Energy and American natural gas pipeline powerhouses Williams Companies and Coastal Corporation.

Alliance Pipeline began in a fashion few pipeliners would appreciate better than Frank McMahon, had he been alive at its conception. Two frustrated natural gas executives met in the Elephant & Castle pub in downtown Calgary in 1992 to mull over the question of who would build new pipeline capacity out of Alberta to relieve the glut of gas that was driving down prices. Gas marketers Glen Perry and Steve Haberl wondered why no one was proposing a direct route from Western Canada to the hub of North America's best gas market, Chicago.

Perry took a napkin, roughed out a map of North America and drew a straight line from Fort St. John, B.C., to Chicago. The moment matched McMahon's 1934 scratching in the sand at Pouce Coupe. In the boozy glow of the Calgary afternoon, the idea looked damn obvious, but in the sober light of day, the two men found it impossible to get anyone else to follow their logic.

Two years passed until Perry drew the sketch again for his boss, Direct Energy president John Lagadin. They enlisted another natural gas man, Jack Crawford, developed a polished presentation and spent a year signing up twenty-two producers as sponsors of the Northern Area Transportation Study. But when they took it to the major pipeline companies, TransCanada PipeLines, NOVA and Westcoast, they got a stony reaction from old pipeline hands who said the producers had no idea how to go about the pipeline business. "When you fail," Perry was told, "you'll finally understand how tough pipelining really is." But Perry, Crawford and Lagadin thought that the pipeline companies had missed the point—the producers were looking for help from people with experience running pipelines. When it wasn't forthcoming, they plowed ahead anyway.

The Alliance Pipeline project, as it was now called, had powerful backing from major producers including Alberta Energy, Chevron, Crestar, Gulf Canada, Pan Canadian Petroleum and Petro-Canada. These companies had deep pockets and a vision for post-regulation pipelines that would be pure merchant operations, closer in their temperament and style to independent gas producers than to fat-cat pipelines.

Alliance was really two projects. The first is the regulated 3687-kilometre, high-pressure 36-inch pipeline from northeast British Columbia through Alberta and across the continent in almost a straight line to North America's biggest gas hub at Chicago. This portion includes, in addition, 698 km of gas-gathering laterals and a total of twenty-two compression stations. At start-up it had 1.325 billion cubic feet per day of capacity.

The second project is the unregulated gas liquids extraction and upgrading plant, Aux Sable Liquid Products, at the eastern terminus. The Aux Sable plant strips 70,000 barrels per day of ethane and other gas liquids from the methane. The pipeline and liquids plants are the most advanced and efficient on the continent, using half the energy to do the same work compared to older systems. Alliance is the competitive alternative to the west-to-east Canadian transcontinental gas transportation and liquids system controlled by TransCanada PipeLines. By breaking up TransCanada's hegemony, Alliance has transformed the Canadian gas industry.

"There were a lot of doubting Thomases, including most of the executives at Westcoast, that the Alliance producer group would hold together long enough to actually get this project off the drawing boards," recalled Mike Stewart. The producers continued with the process of creating a new megaproject. Engineering, financing, market evaluations, right-of-way selection all moved quietly but relentlessly ahead. When TransCanada and NOVA, who wanted all gas liquids to be extracted in Western Canada for the regional petrochemical industry, got into a fierce fight with producers over the issue, it just confirmed to the Alliance sponsors that they needed to break the TransCanada-NOVA west-east transportation monopoly. Alliance formally announced in June 1996 that, subject to regulatory approvals, it would build the high-pressure natural gas pipeline. In December that year, it started to file regulatory applications.

TransCanada and NOVA refused to take Alliance seriously until it was too late, then tried unsuccessfully to block the project at the National Energy Board. Westcoast came at the issue differently. It had experience in and an institutional memory of being a gas producer. It understood both sides of the pipeline conflict from personal experience. Mike Phelps had built an independent producer; Mike Stewart had marketed sulphur, gas and oil as an employee of producers.

Stewart said, "I guess it was February of 1997 that we started to share the view at Westcoast that Alliance was gaining momentum. It was close to achieving critical mass, and if it went ahead it was going to take a significant portion of its gas from northeastern B.C. where our assets are located. We took the strategic decision to be part of it." In September 1997, Westcoast acquired an 11 percent equity interest in Alliance, increasing that to 14.5 percent in February 1998 and 23.6 percent in December 1998, a few days after the National Energy Board gave Canadian approval to the project.

In January 1999, the upstart pipeline received its final U.S. approvals to build a $5 billion, 3687-km pipeline with a launch capacity of 1325 billion cubic feet of gas per day. Construction kicked off on the right-of-way in February 1999 and on the Aux Sables gas liquids plant in Illinois in March. Through the summer of 1999, it was the biggest construction project in North America. In November 1999, Alliance celebrated completion of the cross-border tie-in weld. The pipeline was completed, tested and put into service in December 2000.

Mike Phelps summed up the whole project: "Alliance frightened and delighted me when the founding group got together. Its sponsors said to the producers, 'Instead of being locked into northeast B.C., with only one option to send your gas down Westcoast to the Pacific Northwest, what we really want to do is get to Chicago, the Midwest, the Ohio River Valley with all those big cities. Competing against coal and oil with the pollution problem, natural gas will be the premium fuel. All we need is an outlet across Alberta.' It wasn't a threat to us, with plenty of market in California, so we bought a chunk, then another piece, and then another." Alliance, as a $5 billion project had plenty of action to go around, particularly for Westcoast, because it was already into the Maritimes & Northwest project and Mexico. It was telling that Westcoast increased its equity interest in the project, emerging as the second-largest member of the consortium joint venture.

WESTCOAST'S $429 MILLION equity investment in Alliance set the stage for the company to partner up with Enbridge Energy and MCN to build the Vector Pipeline to take gas from the Alliance terminal in Illinois up to Dawn, Ontario, where Union Gas has a massive storage complex. Vector would move one billion cubic feet per day, some taken from Alliance and some from other sources. It went into service in December 2000, within a few days of the Alliance line. The two projects in tandem will have created an integrated alternative to shipping from Western Canada to Ontario through the TransCanada system.

If Westcoast Energy's $800 million Millennium Project, a proposal to link Westcoast's Ontario system to New York City, succeeds in getting market support, it will complete a natural gas transit link between northeast B.C. and the Liard Basin in the Yukon and Northwest Territories in the Canadian all the way to the America's Atlantic coast.

17

Mexico

We have been made to feel welcome in Mexico. The risk is about the same as in British Columbia under the NDP government. And of all the people I've met in this business, I'd rather have dinner with the Mexicans.
MICHAEL PHELPS

There are two things to remember about doing oil and gas business in Mexico. The first is that it still has highwaymen—bandits who will steal your wallet and your car on a good day, or kill you on a bad one. "You can't make it back to town," is the advice handed out to rural visitors in the late afternoon, "you had better stay over." The second thing to remember is that Petroleos Mexicanos or PEMEX, the state petroleum monopoly, for many years didn't want outsiders on its patch. So insular and suspicious were its managers that it resisted incredible pressure to join the OPEC cartel in the 1970s and 1980s. As a major wealth generator, the state-run enterprise is also vulnerable to the worst of Mexican political cronyism and corruption. The sense of isolation, of being an alien in a strange land, can be close to oppressive at the base for Westcoast Energy's Mexican projects—a remote place called Ciudad del Carmen, 900 km due east of Mexico City as the pelican flies, located with its face to the blue waters of the Bahia de Campeche and its back to the jungle of the Yucatan Peninsula, home to wild jaguars, cursing parrots and blue iguanas. This is a poor, remote rural backwater whose people earn a living by farming the land and fishing for shrimps in the bay.

The Gulf of Mexico's oil and gas reserves are key to the future of the newest member of North America's free trade zone. They are both enormous and comparatively costly and difficult to develop; to do so, Mexico needed foreign capital and—more vital—imported know-how. Following the conclusion of the North American Free Trade Agreement, or NAFTA, Westcoast became one of several Canadian and American companies evaluating the business potential of Mexico's large oil and gas reserves, which were controlled by the government and developed by the country's state petroleum company, Petroleos Mexicanos. Mexico, however, needs foreign capital and—more important—imported know-how. The introduction of private entrepreneurs and the transition from a completely state-controlled energy sector to a mix of private and public investment and operations is proving painful. Powerful Mexican politicians with an entrenched stake in the status quo are resisting. But the country's expanding natural gas development is happening so fast that PEMEX has surrendered its monopoly over natural gas transportation and distribution in order to get international expertise and financing to build and operate new infrastructure. As Mexico opened up domestic competition to national and foreign businesses, it also joined what Adrian Lajous Vargas, head of PEMEX, called "a natural gas market that stretches from the Yukon to the Yucatan." Not surprisingly, with its two Mexican projects, that also grew to be the extent of Westcoast Energy's continental reach.

It soon became clear that the first opportunities would be in developing energy infrastructure rather than investing directly in oil, gas and power production. Westcoast's initial crack at Mexican business, which came in 1998, wasn't an investment but a consulting services contract, in partnership with U.S. giant El Paso Energy International, with PEMEX to modernize its natural gas control system. Previously, because PEMEX owned the production, the pipelines and the distribution and end-use facilities, there had been no need for gas measurement; the ownership of the gas never changed. However, Mexican energy reforms were evolving and, eventually, other companies would take ownership of natural gas at various points within the infrastructure. Technical experts from Westcoast and El Paso teamed up to work with the Mexicans to install a computerized measurement system.

Westcoast had its eye on pipelines and distribution but failed to win the bidding when PEMEX auctioned off three gas distribution franchises. Westcoast's bids were too low but flagged the company as a serious player.

In 1997, shortly after the auction, Harvey Permack, a long-time Westcoast executive, received a call from an acquaintance, who knew of the Vancouver Island pipeline project, Westcoast's interest in electric power and its initiatives in Mexico. "The call came out of the blue," said executive vice-president Michael Stewart, who, as president of both Westcoast International and Westcoast Power, had led the company's international business development since 1997.

British industrial gas giant BOC Gases asked Westcoast to join it to bid on a plant to manufacture and supply nitrogen for an enhanced oil recovery project at the Cantarell offshore oil field in the Bay of Campeche in the Gulf of Mexico. It was a huge opportunity: Cantarell produced a million barrels a day and ranked as the sixth-largest field in the world. The competition was intense, with Canada's NOVA Corporation in partnership with Marathon Oil and Houston-based Enron Energy teamed up with Shell Oil and the Japanese financing house Mitsubishi.

To maintain production rates, PEMEX needed to re-pressure the oil reservoir. The company had been flaring natural gas that was being produced in association with the oil. However, as Mexico increased its use of natural gas, it became too valuable to flare, so PEMEX was looking for an alternative and put out a request for proposals on the economic terms and technical conditions to produce nitrogen as a re-injection gas. This was the kind of business BOC Gases was doing all over the world. Westcoast was asked to join the consortium to provide the expertise needed to pipeline the nitrogen gas to re-injection wells, and to build and operate the 220-megawatt gas co-generation power plant required to produce the nitrogen. The BOC Gases group included engineering giant Flour Daniel (partnered with ICA, a Mexican construction firm), Westcoast, Japanese trading company Marubeni, and Linde AG, a German heavy industrial equipment firm.

When it came to the showdown, the BOC Gases group won with a U.S. $1 billion (Can. $1.5 billion) bid that was a mere U.S. $27 million lower than the highest of the three bids. Westcoast Energy is a 20 percent member of the Cantarell Nitrogen Project consortium that was finally awarded the fifteen-year contract to build, own and operate the $1.5 billion nitrogen production and delivery complex near Ciudad del Carmen, in the State of Campeche, Mexico. Westcoast Energy contributed to project design and jointly operates the facilities, which went in service in 2000. Cantarell is the world's largest nitrogen production and delivery plant. Four air separation units produce 1.2 billion cubic feet of high-quality nitrogen at an onshore plant, then deliver it through a high-pressure onshore and offshore pipeline system to a PEMEX injection platform.

Constructing the Cantarell facility became Westcoast Energy's graduate school seminar in doing business in Mexico. PEMEX was to provide a permitted site and Westcoast assumed that the contractor could start clearing the site on day one. It wasn't quite as easy as that. PEMEX had treated landowners badly and local resentment ran high over right-of-way and plant site payments. Westcoast was confronted with demands from the local farmers and shrimp fishers. They presented a petition with one hundred demands, including a 21-mile canal and fishing access across the project site. "This was not part of our game plan," deadpans Michael Stewart.

The company didn't succumb to the local demands but tried to work along with the community. "We were going to be there for at least fifteen years and were going to have to contribute something back. But we couldn't be held to ransom and we didn't want to be in a position of paying out money for no particular purpose," says Stewart.

The shrimp fishermen were worried about the environmental effects of the plant discharge water and about how laying the offshore pipeline would disrupt their fishing activities. These were typical demands one expects from people where construction is involved and right-of-way has been acquired. Really what Westcoast faced was the fallout from PEMEX's modus operandi in acquiring land, and from the expectation that it was a rich foreign company ready to pay out to get ahead. More importantly, says Stewart, the local communities are not very well off in an economic sense, and the people are looking to better their lives. Most of the four thousand construction jobs went to skilled workers brought in for the project, but there were local spinoffs.

Westcoast worked with the community to set up a foundation to promote local economic opportunities, including the creation of a credit union to establish a source of financing for small business. With Westcoast taking the lead, the project partners funded a medical clinic and worked with the Canadian International Development Agency to secure a $500,000 CIDA contribution to local projects.

Cantarell was also the first major industrial project to go through a formal and complete environmental review process in Mexico. It broke new ground, applying environmental standards not unlike Canadian requirements in terms of plant construction. The project was given an environmental permit that had forty-three conditions attached, and it took considerable time and money to start construction. The project lost four months off its schedule, and costs jumped by nearly $10 million. The partners did significantly more background environmental studies and impact assessments. The community wanted to protect birds from electrocution on the power lines, and to preserve wildlife habitat. The shrimp fishers questioned the discharge water's impact on fish and marine life. A noise mitigation study and program was required.

"The level of environmental standards in this plant are world class, and neither the company nor the consortium are trying to do anything to get around those. But the basis on which we bid for the project is not the basis upon which it was completed. Some of what PEMEX said was there, was not there. We had to deal with that and it involved extra time and money," Michael Stewart explained.

Security became a big issue because of local political unrest. The plant site roads were blockaded ten times during construction and protesters entered the site. In early June 2000, the entire plant site was sealed off and Westcoast airlifted people in and out by helicopter while running a performance test.

Finally, at the urging of the consortium, the Mexican government declared the plant a strategic installation. The government assigned a company of marines to be deployed at the gate. The partners built a barracks for the marines at a cost of half a million dollars and paid the bill to feed and house the troops. But the troops were an effective deterrent to blockades and they enhanced the perimeter security. "We were sitting there with over $900 million in hardware and very little protection for it," Stewart recounted. "Security became much more of an issue than we had contemplated."

The U.S. $1 billion Cantarell financing was one of the most difficult transactions that Westcoast had dealt with since Frank McMahon gerrymandered the backing for the British Columbia main line. As a matter of policy, Westcoast doesn't expose its own credit on international projects after the construction phase. Each facility had to finance itself. Cantarell was a sound project on its financial merits, but the Third World and particularly the Latin countries were under a major financial cloud. The value of the Mexican peso was fluctuating, so the quality of the credit of Mexico was under suspicion. The Asian financial flu had devastated Japan, Indonesia was in a meltdown, and there had been serious problems in several South American countries.

The idea had always been to bring in Japanese Export Import Bank, which is like Canada's Export Development Corporation, as a financial partner. The Westcoast group interviewed a variety of investment banks with the intention of tapping international capital markets. "When the world financial climate worsened in 1997, there was no way to get to the capital markets and we had to reassess our options," Graham Wilson pointed out. Eventually the project partners assembled a commercial bank consortium. It was a challenge, but Citibank of New York emerged as the lead facilitator because it had a particularly good relationship with the Japanese Export Import Bank and was the only non-Japanese financial institution it found acceptable as an agent bank.

"It was easy dealing with Citibank and the commercial banks," Wilson said. They came to the table fairly quickly and what they required was reasonable. What was more difficult was getting other investment banks to come in. They were very concerned about their Mexico limits. The price of oil was down and this project was a transaction with PEMEX. Therefore it depended on the financial strength of Mexico and PEMEX itself."

The investment bankers were also fearful about the political instability created by the Zapatista National Liberation Army guerrillas who had been active in Chiapas province since 1994 and were opposed to NAFTA and foreign investment. "Even though the Cantarell nitrogen project wasn't near where those guerrillas were operating, nobody wanted to go anywhere near those types of issues and exposures," said Wilson. "So we did a lot of reassuring and we used every connection we had in the financial market. We put together a package with Canada's Export Development Corporation, Citibank and Fiji Bank. We tapped the market for every penny that was out there. And we were still short about U.S. $360 million."

The partners then brought in the Japanese Export Import Bank from the sidelines, and it was difficult to deal with. It had a very substantial exposure to PEMEX already and was looking at the Cantarell financing as part of its total aggregate exposure. It also had a deliberate style of working and assessing the credit. Lengthy negotiations ensued, jumping around between New York, Mexico City and Tokyo. Assembling the deal took more than a year longer than originally planned and the Japanese rates were slightly higher than anticipated.

However, even though the partners had squeezed every dollar out of the commercial banks, and pushed the Japanese as far as they would go, they were still short of the money required. Under the circumstances, however, they felt fortunate to have ended up with over 60 percent of the funds required through project financing with the banks and institutions.

"The key thing we had achieved was that in all these transactions the lenders had recourse only to the project and not to Westcoast Energy," Wilson noted.

Getting Cantarell built was the next challenge; even though Westcoast knew that the business milieu was challenging, there were unexpected twists and turns. Westcoast participated in the first environmental review process in the history of Mexican energy development and found that the social revolution became part of the process. "We were in the midst of that revolution. Local farmers and fishermen and Natives had a voice in development for the first time in forty years. Although that was nothing more than the usual aggravations, we were several months behind waiting for approvals and, in North American terms, time is money," Mike Phelps explained.

Doing business in Mexico was a long-term commitment for Westcoast, even with the frustrations of having to deal with a different economic culture. "We were the first Canadian company, and one of the first worldwide companies to make a direct equity investment in a facility providing service to adjacent PEMEX's oil and gas operation. PEMEX had been very reluctant to see that happen. Its nationalization has been celebrated for many years as a defining political event on the road to independence," Phelps noted.

Westcoast didn't breeze in and out of foreign countries cherry-picking for opportunities, but spent time developing its understanding, creating relationships and putting down a foundation. Mexico illustrated this point in three ways. The business culture is much less intense than that of Canada or the U.S., demanding patience and flexibility. The political establishment is tightly interwoven, mysterious and inaccessible, with many internal political conflicts and unresolved rivalries. Westcoast succeeded, said both Phelps and Stewart, because of the relationships it developed with senior people in government and PEMEX, and the reliance it placed on Mexicans to assist the company in solving problems.

Outside companies have to temper their expectations to Mexican circumstances. The Campeche project didn't run as smoothly as Cantarell. The prime contractor got into financial difficulty, and this slowed down completion.

Westcoast had $550 million invested in the Mexican energy sector. At the company's 1999 annual general meeting, Michael Phelps told shareholders, "We see Mexico as a full member of an integrated North American energy economy. Our aggressive development and considerable capital investment in Mexico will extend our reach in a growing continental energy market. We believe that the risks of investing in an emerging market such as Mexico are manageable. We anticipate substantial and appropriate returns over the life of our projects. We also anticipate follow-on projects in this area, once our two large projects are closer to completion."

In July 2000, Mexican voters shook up the political system by ending the seventy-one-year tenure of the left-of-centre Institutional Revolutionary Party (PRI) and electing Guanajuato provincial governor Vincente Fox of the Catholic National Action Party (PAN).

Fox is an educated politician with degrees from the Iberian Americano University in Mexico City and Harvard. He is a former business executive with Coca-Cola, and a free trader who wants a seamless North America from the Beaufort Sea to the Panama Canal. He promised to loosen the stranglehold that PEMEX maintained on petroleum development and, on visits to Washington, Ottawa and Toronto, courted more foreign investment, especially from the rest of North America, in Mexico and its petroleum sector. He indicated that the petrochemical sector would be opened for business. PEMEX, however, has a Byzantine control over oil and gas matters and would not be easily persuaded to cooperate with the new president. PEMEX generates 30 percent of Mexico's federal revenues, and is credited with engineering, with a cooperative OPEC, the sustained run-up in oil prices in the year 2000.

Westcoast Energy proved its knack for Mexican diplomacy with the federal government, PEMEX and local interests. The petroleum sector was starved for foreign expertise and needed outside capital. European and American companies operating in Mexico put Westcoast on their short list of desirable project partners, and Mexico remained the preferred arena for Westcoast to expand its investments beyond Canada and the United States.

18

The Canadian North

We see the North differently because that is where we are from.
MICHAEL PHELPS

The platform from which Westcoast Energy looked at North America's Arctic gas opportunity was different from that of any other gas pipeline company in the twenty-first century, because Westcoast lived in the North. More important, it felt a northern identity deep in its corporate bones. The company had operated on the threshold of the sub-arctic for forty years. Its Fort Nelson gas processing complex, at 58 degrees, 48 minutes north latitude, is just seventy-five miles south of the sixtieth parallel.

Its pipeline system crossed the border to the Yukon in the 1960s and to Fort Liard in the Northwest Territories in the 1990s, and it became the first Canadian gas pipeline moving sub-arctic gas to continental markets. In the process of completing these projects, Westcoast developed and maintained successful and strong relationships with Cree and Dene aboriginal communities in British Columbia's controversy-prone Treaty 8 and Western Arctic land-claim areas. Westcoast's sense of the North played a role in the complex motivations that led it to participate in the creation of the Muskwa-Kechika Management Area in British Columbia, a wilderness protection zone the northern boundary of which is contiguous with the sub-arctic.

The heart and soul of Westcoast's northern vision, however, was in its multi-million-dollar, and more than thirty-year-long involvement in the challenge of building big-inch pipelines to bring Arctic gas south. Frank McMahon hired Ed Phillips to put Westcoast Transmission in the lead position for his scheme of two projects: one from Alaska's North Slopes and one from the Canadian Mackenzie River delta. It is evidence of his prescience that, although he was impossibly ahead of his time and the project had no more weight than his 1954 offer to build a gas pipeline from Alberta to Toronto, his concept of two pipelines from the Western Arctic became a hot and heavy competition, after 2000, to see which one got constructed first.

Ed Phillips regarded it as a career achievement that Westcoast, under his tutelage, participated in the Arctic Natural Gas Transportation System of the 1970s and became a partner in Foothills Pipe Lines, which pre-built two legs of the system to carry Alberta and B.C. gas into the United States. Foothills held permits protected by a Canada–U.S. treaty to complete the line from its west-central Alberta terminus north through the Yukon and Alaska to Prudhoe Bay. It had completed rights-of-way, design and engineering studies, and environmental and community assessments—costly predicates for an actual project.

And an Arctic natural gas pipeline megaproject was a challenge suited to the company's temperament—from Frank McMahon's wilderness pipeline in 1955 to Michael Phelps's Campeche and Cantarell Mexican plants, Westcoast was at its best when the job was impossible. "Pipelines from the Arctic will go through several economic cycles from the time a commitment is made until the gas starts to flow—over seven or eight years," Phelps said. "The issue is financing it, and that depends on risk."

When Westcoast first pursued the Arctic opportunity, regulated pipelines controlled the agenda and drove the projects. In 2002, the producers drive the decision—it is their gas, and the pipeline company is the intermediary expert who can build and run the infrastructure. Nevertheless, Westcoast maintained the role of project advocate and champion, and worked with the Alaska producers to find a way to make the project work. In spring 2002, with natural gas prices at 100 percent more than they had been three years previously but not high enough to make a project economic, the major Alaska producers continued to work with Westcoast and its competitors to find and dive through the inevitable window of opportunity.

The politics had changed, too—particularly with respect to aboriginal relations. During Ed Phillips's attempt to get the Alaska pipeline built, First Nations were consulted and their sensitivities about environmental impacts and the rough transition from isolated, land-based communities to modern economies considered. By 2000, Canada's original peoples were emerging as legal and political First Nations, with a myriad of treaty negotiations underway to define their future, and considerable leverage, perhaps even the decisive say, on the terms of development.

If, as Michael Phelps believed, the CEO's job is to set corporate strategy, find the people who can execute it and maintain the external relationships needed to achieve it while the troops get on with the task, then a North American northern pipeline depends on the CEO for its success in spite of the political and economic barriers. Getting that right for Westcoast was one of Phelps's achievements, though the pipeline remained on the draftsman's board.

WESTCOAST ENERGY EXISTS in a three-way relationship with the land and the aboriginal communities in northern British Columbia and the Northwest Territories. The physical pipeline and the indigenous people share the same landscape; they have shaped it, and it has shaped them. And the land—in all its magnificence—has been the stage on which they have worked out their relationship, based on shared values that are the product of what they have taught one another.

A pipeline leaves only a faint trace on the land—the thin scar of the right-of-way, which might be thousands of miles long but is never more than a few yards wide. Westcoast Transmission changed the nature of the Peace River country by bringing with the pipeline an entire new resource industry: oil and gas exploration, drilling and production. And the roads, well sites and processing plants required hundreds of acres of surface space, creating a new skyline. Constructing the Westcoast pipeline was a defining act of British Columbia's resource development. The people who came to operate this new industry required homes, schools, churches, parks, hockey rinks, shops and stores. Pockets of urban development spread across the land. The cumulative impact of the pipeline over more than forty years has been to permanently change both the physical environment and the attitude of its First Nations.

No account of the life of the newcomers—the letters they penned, the oral interviews they gave, the books they wrote—can miss that, in turn, the land changed them. There are the obvious ways: the hunting and fishing and camping that created a community of outdoorsmen and outdoorswomen. There are the less obvious ways: the characteristic self-deprecating humour, the self-reliance, courtesy and gentleness that the pipeliners acquired when they stayed. To the outside observer, these traits have made the pipeliners and roughnecks more like their aboriginal neighbours and less like the people they left behind in southern cities and towns.

In terms of environmental programs, resource companies like Westcoast have a duty of compliance: there are regulations and standards, and meeting them is hard, expensive work, requiring tight management systems, but it is not optional. Pipeliners always try to outdo not just each other but also the producers and the refineries and chemical plants at either end of the pipeline. Westcoast makes a point of exceeding requirements.

The principles and values behind environmental compliance, protection and mitigation have become bred in the bone, an ingrained part of Westcoast's corporate culture. Wayne Soper, senior vice-president of external relations, said the culture of environmentalism is "part of the company's DNA," and "in the tummies" of senior executives and the field crews alike. "It's more than being Green—it's knowing that you live and work in an organic world," says Soper. "And it is inseparable from the company's relationship with the community. Communities are one leg of the stool of sustainable development, so we are removing government as the proxy for the community and dealing directly with our public."

At the heart of this attitude and outlook, says Soper, is the company's relationship with aboriginals. Part of it takes the form of corporate contributions to hospitals and schools. The investments recognize that the community is part of the operation's bread and butter. Part of it is sulphur recovery at processing plants—removing hydrogen sulphide from plant emissions and the air the communities breathe. A big part of it is engaging the community in the planning for future facilities. The most important aspect is that the playbook comes from the direct relationship between company and community—and not from the desk of a regulator trying to find common-denominator standards for compliance and mitigation. "We align ourselves to be part of the solution, rather than being the problem," Soper contends.

How that came about is a matter of geography. Westcoast met the aboriginal community and its environmental responsibilities at the same place: the first right-of-way in the most beautiful and environmentally vulnerable portions of British Columbia. Where the right-of-way crossed Native reserves, Westcoast mastered the politics of 1950s-style dealings with the federal Department of Indian Affairs and learned to reach around the paternalistic bureaucrats to establish working relationships on the ground. The common ground was jobs, and employment became a cornerstone of the relationship. Dozens of aboriginal men worked for the company; some made careers of it. The relationship had its critics, none tougher than retired CEO Ed Phillips, who has said in retrospect that the company wasn't very successful for many years in creating economic opportunity. But the quality of the relationship, from the aboriginal perspective, was more than had been expected. One could argue that Westcoast looked good because others were so bad. However, there's evidence of a cultural difference between Westcoast's philosophy and practices and those of other companies.

The secret of the three-way relationship between the company, the land and the aboriginal community is most evident at Fort Nelson, the second-most-important town on Westcoast's British Columbia map after Fort St. John. It is the hub of a major complex of gas fields and processing plants, as well as the lift-off point of the company's sub-arctic pipeline over the sixtieth parallel of latitude to Pointed Mountain and the Liard Basin. And it is the symbolic meeting place of the land, the aboriginal community and Westcoast Energy.

Fort Nelson, located about 50 miles east of the Rocky Mountain foothills, is on the same latitude as Skagway, Alaska and Stockholm, Norway. It is situated on a gently rolling plateau, an extension of the Great Interior Plains. Wooded slopes rise steeply from the east until they level out into this plateau. It is a place where the Muskwa, Prophet and Sikanni Chief Rivers converge to form the Fort Nelson River, which flows into the Liard and from there to the Mackenzie out into the Arctic Ocean.

A large part of the area is muskeg, the swamp of spongy ground covered with moss. It is part of the Great Northern Continental Boreal Forest that extends from Newfoundland to the Rocky Mountains. The trees are mostly coniferous, with spruce in abundance and a few pines as well as some birch and balsam, poplar and willow. Grass is scarce, but there is much wild clover, wild roses, fireweed and dandelions.

This is trapping country. The woods are still full of beavers, wolves, weasels, foxes, lynx, minks, muskrats and marten, as well as deer and caribou and grizzly bears. Geese and cranes fly overhead, as do swallows, starlings, robins, great grey owls, rock doves and chickadees. In the summer the place is abuzz with mosquitoes and blackflies. It is still home to many aboriginal groups, including the Beaver, the Sicanee, the Nahanni and the Dogrib. In 1775, the Slave Band moved here from the Great Slave Lake area, driven east by the Cree, who were the first to acquire firearms. The term "Slave" was intended to characterize the great humility of these people as compared to more warlike tribes.

Fort Nelson was founded in 1805 by the North West Company, named after Admiral Nelson, who, that year became the hero of the Battle of Trafalgar. The first white men to arrive were the Nor'West fur traders who came by way of the Mackenzie River to Fort Simpson and from there up to the Liard and Fort Nelson Rivers. In 1865, it became a

Hudson's Bay Company post. The fort was destroyed by fire on one occasion and flooded out on another, but the community stayed and rebuilt. There are those who remember their fathers and grandfathers pulling loaded HBC York boats up river with long ropes that reached to shore. The boats remained in service for other purposes until after the Second World War, and there are still men around who remember doing the gruelling work of dragging those boats against the current in all manner of rain, sleet and blackflies, regardless of whether the river bank was rock or muskeg.

The local airport was built in 1942 as part of the staging procedure to build the Alaska Highway. After the war, Fort Nelson declined; its only outside economic support came from the Canadian military. When Westcoast arrived on the scene in the late 1950s, even that had all but disappeared. What was left was a struggling Aboriginal-majority community, and Westcoast had a tough time getting its people to stay. But the company pitched in, through a unique voluntary taxation system, to build up community infrastructure, creating jobs and businesses. It hired more Slavey people and dispensed more contracts to Slavey businesses here than at any other place along the line. The two communities were thrown together in the splendid isolation of the Canadian wilderness, and through the osmosis of human relationships built a partnership of common interest.

It's a mark of the quality of that relationship that the two men who served as chiefs since Westcoast arrived both picked up paycheques. George Behn, chief since 1950 except for the period 1970–76, organized line-clearing and reclamation crews. The company wanted him full-time, but being chief was his day job. Harry Dickie, chief from 1970 to 1976, worked as a plant caretaker for a few years, but he didn't much care for indoor tasks and went back to his trap lines. Dickie arrived in Fort Nelson from Vermilion Forks in 1933 with a packhorse train, wanted to go back but stayed. He became a member of the school board and director of Takima Forest Products in which the Slave band holds a 25 percent interest. He speaks four languages—Cree, Slave, French and English. As a judge of Westcoast's environmental or aboriginal relations programs and polices, he is far more qualified than any bureaucrat or activist, and much tougher to fool or impress. Westcoast did the only sensible thing they could do with men like Behn and Dickie: they enlisted them as partners in managing the community and managing the land.

One of the lessons Westcoast Energy learned in doing business with communities such as Fort Nelson is that it was judged by what it did, not what it said. While community leaders are inclined to regard four-colour brochures as a token of respect, what matters is action. On October 8, 1997, the British Columbia government created the 4.4-million-hectare Muskwa-Kechika land-use protection area in the northern Rocky Mountains. Larger than Switzerland—about the size of Nova Scotia—the land-use plan protects 1.17 million hectares of the area, and sets aside 3.24 million hectares, or 77 percent of the land, for special management, including strictly controlled resource development. The guidelines for land use had been worked out in a consensus of scientists, environmental activists, First Nations and the oil and gas industry—with the Canadian Association of Petroleum Producers and Westcoast Energy playing a leading role. Two oil companies, Amoco Canada and Talisman Energy, surrendered undrilled oil and gas drilling leases in which they'd invested $500,000. Westcoast and Petro-Canada contributed $100,000 seed money to a fund for research into how oil and gas operations should take place in selected special management areas.

"It was," said Michael Phelps, "one of the best day's work Westcoast ever did." The Muskwa-Kechika is the largest remaining big game area in the world, outside of the Serengeti Plain in Tanzania. "The Serengeti," said Phelps, "is black rhinoceroses, elephants, zebras, gazelles, cheetahs, lions and leopards. The Muskwa-Kechika is grizzlies and black bears, moose and elk, deer, cougars, mountain lions. The lakes, mountains, rivers and forests are vast. I have travelled there on horseback and it is matchless. Its vastness puts Westcoast's continental reach into proper perspective. We say a pipeline can be invisible and has a light footprint, and the Muskwa-Kechika challenges us to live up to that claim. It is also our bridge between our past in the remoteness of northeast B.C. and any future we will find in the North. I doubt Westcoast ever did anything more important than participating in the Muskwa-Kechika."

Westcoast Energy's leadership in creating the management area earned it the 1999 Ethics in Action Award from a coalition of Canada's leading corporate ethics organizations. What the Muskwa-Kechika conservation represented was the sum of the lessons Westcoast had learned operating in the B.C. Interior, and in communities like Fort Nelson. In a word, the lesson is balance—the balance of preserving wilderness integrity, safeguarding jobs and managing resource development.

WITH THE ALLIANCE, Vector and Maritimes & Northeast pipelines all completed, Westcoast Energy turned its attention to the Arctic, back to its Foothills proposal to develop the Alaska natural gas pipeline and the tangential opportunity to build a pipeline out of the Mackenzie River delta. Michael Stewart, executive vice-president of business development, and president of its international and electric power operating arms, was assigned the task of developing the project. There Westcoast faced another period in which pipeline planning is mostly politics. But the issues are much different than in the last round of northern pipeline politics, when environmental and land-claim conflicts brought oil and gas commercialization north of the sixtieth parallel to a dramatic halt.

Between the summer of 1960 and the autumn of 1988, oil and gas exploration in the Mackenzie Delta, the Beaufort Sea and the Arctic Islands resulted in the discovery of more than three dozen major natural gas fields and an estimated 100 trillion cubic feet of gas, spread out across a 1550-mile frontier. Briefly, in the oil and gas price boom from 1973 to 1981, it looked as if developing those reserves might be profitable. Parallel discoveries in Alaska's Prudhoe Bay led to the planning of three gas pipeline corridors, and the sketching out of two major liquefied natural gas (LNG) terminals. The proposed pipeline rights-of-way ran from Alaska through the Yukon and northeast British Columbia to Alberta, from the Mackenzie Delta to Alberta and from the Arctic Islands south past Hudson Bay to Montreal. The LNG proposals involved shipments by sea from the Arctic Islands to Europe and from the Western Arctic to Japan. Similar oil exploitation schemes were afoot. Developers wrapped their projects in the Canadian flag—oil and gas had to come out of the North under Canadian jurisdiction to protect national sovereignty.

The commercial development of northern oil and gas ran into two land mines. The first was resistance from Inuit and Cree aboriginal peoples unprepared for the economic development of time-honoured Native lands, and the inevitable end of their hunter-gatherer lifestyle. The second was from environmentalists, who feared the desecration of the North. An alliance of these two working groups successfully foreclosed on development until the collapse of oil and gas prices in the mid-1980s ended commercial interest. By the time the moratorium kicked in, pipeline consortia had spend $250 million in engineering, right-of-way selection, environmental assessments and regulatory expenses. Before the dust settled, however, Westcoast Energy had positioned itself as a 50 percent partner in Foothills Pipe Lines. Westcoast spent $110 million as its share of the cost of a Y-shaped pipeline from Caroline, in central Alberta, with one leg through Saskatchewan to the U.S. border, and the second through British Columbia to Washington State. The pre-build, as the system is called, carries natural gas out of Alberta, earns Westcoast about $10 million a year, and is intended to be the end point of an Alaskan gas pipeline with a possible lateral connection to the Canadian Western Arctic.

Westcoast's commanding advantage in the competition to build the northern gas pipeline is that, through its 50 percent stake in Foothills, it owns a pivotal piece of the regulatory approvals to build its route, which can carry both Alaskan and Mackenzie Delta gas. Those approvals are enshrined in Canadian and American federal legislation, and in a treaty signed by then–Prime Minister Pierre Trudeau and then-President Jimmy Carter. "If you had asked me two years ago about a pipeline from the North in my working career, I would not have bet a nickel on that," Michael Stewart said in the summer of 2000. "Now I am spending a quarter of my time on the opportunity."

Alaska and Canadian Arctic gas will provide a huge chunk of the gas required to fuel the North American economy. And now, Canadian Arctic aboriginal peoples are advocates of development. In the fifteen years since oil and gas development went on hold, the Arctic has seen a major mining boom—diamonds, gold and base metals are being produced from four major new sites. An oil pipeline has been constructed down the Mackenzie Valley from Norman Wells. And gas producers are commercializing major new discoveries in the Liard Basin, a prolific gas area tucked into the southwest corner of the Northwest Territories. The producers have so far discovered 2.2 trillion cubic feet of natural gas in the Liard Basin, and they're certain that they've just scratched the surface. Westcoast Energy operates the first sub-arctic gas pipeline, which runs north from Fort Nelson, crosses the border in the Yukon, and swings into the Northwest Territories to serve the Pointed Mountain gas field.

In January 2000, Westcoast Energy signed a processing contract with the West Liard Valley Producers' Group to develop a gathering system from Pointed Mountain to three major gas wells, each producing 75 million cubic feet per day, and ship the gas to Fort Nelson for processing. The company is also planning construction of gathering lines to three more major wells. In other words, Westcoast Energy has, quietly but effectively, gained control of gas transportation from the first commercial north-of-60 fields.

The bigger prize is a gas line down the Mackenzie Valley. Half a dozen major natural gas exploration and production companies raced back into the Mackenzie Delta in 1999 and 2000 to re-open drilling. It will take at least two and possibly four years to get through the regulatory system—as environmental concerns, land claims and other issues are settled. This time, however, most northern aboriginals, who by law will appoint half the adjudicators on environmental and regulatory panels—want development. And there is an almost irresistible demand from the continental economy—gas marketers and customers who recognize that demand in the lower forty-eight states is growing exponentially. Nobody wants to use coal or nuclear fission to close the gap, and there's little interest in importing more oil if continental natural gas is available. For the first time since the idea was coined that clean-burning natural gas is the bridge fuel from burning oil to the clearer air of the future, the notion that gas is environmentally friendly will pay a pivotal role in a major continental regulatory process.

Mike Phelps was clear on the opportunity: "With the present gas production decline rates in the U.S., and Canada, Mackenzie Delta gas looks more and more prospective each month. It's five to ten years away rather than beyond that time horizon. It needs more drilling but there's gas there in the Delta, onshore mostly. It's a race. It's a big project that will require partners, and the Canadian industry will have to have some kind of understanding how to proceed."

Northern gas pipelines are an immense corporate prize—probably the last one-shot, multi-billion-dollar additions to pipeline companies' asset bases. So the toughest fight in this round of northern development will be between pipeline developers. The Mackenzie Valley project is so large that it will require a consortium to develop it. In the spring of 2002, even the largest of the U.S.–controlled pipelines were leery of taking the risk of going it alone. But it is difficult to imagine a Mackenzie Valley line without the participation of a company of Westcoast's stature.

Westcoast, through its 50 percent interest in Foothills Pipe Lines holds a competitive stake in the Alaska Highway route. Westcoast dominated transportation and processing in the Liard Basin. And it is a likely partner in a Mackenzie Valley gas pipeline consortium. The threat of modal competition—from a renewed interest in LNG shipments to Asia—remained. It was a prospect that would motivate the formation of a broadly based consortium, rather than a real threat to the seemingly inevitable pipeline franchise.

The future of LNG technology—still beset by public safety concerns over the perceived potential for LNG terminal disasters, exacerbated by the September 2001 terrorist attacks in the eastern United States—is as much a Westcoast opportunity as it is a threat to future pipeline projects. As a partner, through Foothills Pipe Lines, in the North Slope

Alaska project, Westcoast is participating in a study examining the possibility of sending Prudhoe Bay gas via a pipeline to either Valdez or Cook Inlet. There it would be liquefied and tankered to terminals in Japan, the United States and elsewhere on the Pacific Rim. The members of the study include Phillips Petroleum, BP, Foothills Pipe Lines (Westcoast Energy and TransCanada PipeLines) and the Japanese trading firm Marubeni.

19

The Cornucopia

> *The opportunities for continued investment in natural gas infrastructure in North America were limited. We had skills and capabilities that would serve us well in emerging energy markets.*
>
> BOB REID

As it spread across North America, Westcoast also chased the lure of global investment. The founding president of Westcoast Energy International, Bob Reid, faced stiff competition—particularly in the Chinese market—from companies in the U.S., Europe and Asia. "We did have some notable accomplishments," he recounts. "We established Westcoast Energy in China, Indonesia and Australia. We were the only Canadian power-developer to land a successful project in China, a natural gas co-generation power plant in Shanghai. Our customer is Number One Iron and Steel; the plant is 50 megawatts, but expandable in the future to 100 megawatts. We are also equal partners with Duke Energy, based in Charlotte, North Carolina, in a very significant power plant on the island of Irian Jaya in Indonesia for a very large copper mine operated by Freeport McMoran."

Additionally, Westcoast Energy International developed a pipeline project in Australia to bring gas from an offshore field in the Bass

Straits of southeastern Australia to the city of Sydney. The company decided in 1998 not to proceed because the projected returns were too slim; and it sold the engineering drawings, permits and other plans to Houston-based Duke Energy for $17 million. Although it had developed a permanent taste for international ventures, especially in electric power generation, Westcoast was finding better opportunities on its own continent.

With the 1999 sale of its Australian pipeline project and its decision to concentrate more heavily on Mexico, Westcoast Energy pulled back from investments and operations outside North America. The company kept a watching brief on international opportunities. With its capital and expertise occupied in increasing the density of its interests in its prime North American geographical regions, it wasn't obvious that the company could strike another international deal at any time. Westcoast's people were on the prowl in Europe, South America and Asia for openings that fit its pocketbook, experience, skill and appetite for investments that will provide rates of return higher than its core, utility-style assets.

IN EVERY ONE of the quadrants of focus on Westcoast Energy's North American map, there were opportunities for gas-fired electric power generation projects. Westcoast amassed an impressive portfolio of power operations on the continent with interests in power plants from New Brunswick to Vancouver Island.

Westcoast's power projects included 450 megawatts of capacity in two Pacific Rim projects, in the People's Republic of China and in Indonesia. These interests were held with blue-chip partners and provided power to equally blue-chip industrial customers.

Westcoast's first Pacific Rim investment was its 42.8 percent interest in the $800 million, 388 megawatt multi-fuel PT Puncakjaya power plant on the island of Irian Jaya, Indonesia. In 1994, the joint venture purchased the 193 megawatt hydro and diesel generators and in 1997, the 195-megawatt coal-fired power plant and transmission lines that provide electricity to PF Freeport, Indonesia's copper and gold mine on the island. It is the largest copper producer in the world and is owned by New Orleans–based Freeport McMoran. Duke Energy was an equal 42.8 percent partner and PT Prasarana Nusantara holding the remaining 14.4 percent.

The company's offshore outposts included the U.S. $50 million, 50-megawatt Wei Gang plant in Shanghai, People's Republic of China, for the nation's largest steel plant. This is a conventional coal-fired power plant with a co-generation feature that employs waste heat from the blast furnaces to generate additional electricity, reducing emissions as well as improving the economics of the plant.

Westcoast had a preferred technology—gas-fired co-generation and combined cycle plants—in the sector of the North American electric industry generation that is the most deregulated. In the one remaining area of regulation—the environment—natural gas is the fuel of preference. According to Mike Phelps, "We had opportunities, whether we merged or bought or built." Private gas-fuelled electrical plants have short lead times and easy financing hurdles to cross, compared to their hydro, nuclear and coal-fired megaproject competitors.

Two of its North American power plants take natural gas fuel from Westcoast-operated equity-interest pipelines. Island Cogeneration CP is a 100 percent-owned, 250-megawatt combined cycle plant near Campbell River, B.C. The $220 million plant went on line in the autumn of 2000 and draws its gas from the Vancouver Island pipeline. It sells its electric power to BC Hydro and its steam heat to Fletcher Challenge's Elk Falls pulp and paper mill.

In Saint John, New Brunswick, Westcoast was the 75 percent owner of the Bayside Power Project, which was developed with Irving Oil Limited to utilize fuel from the Maritimes & Northeast Pipeline. Westcoast converted an oil-burning power generator to natural gas, lifting its capacity from 100 megawatts to 285 megawatts.

Westcoast's final power project was the Frederickson Power Project near Tacoma, Washington. Westcoast and EPCOR, Edmonton's municipal electric utility, purchased the plant in 2000 when construction was partly completed. Engage Energy joined the $166 million, 250-megawatt project as the energy services manager. Westcoast owned 60 percent of the project, which will be completed in 2002.

Electric power generation is an opportunity with frustrating constraints. Growth of the business takes place in very small increments. Rates of return are negotiated with customers and must compete with big utilities—so profits are not as lucrative as in other lines of business. More projects are abandoned at some stage of planning than are developed. And there are scores of big electric utilities, natural gas pipeline companies and even gas producers who want a piece of the action, so every potential project is subject to fierce bidding. Preparing a competitive offer costs a lot of money, which is written off if the proposal fails. So electric power project development is one of the toughest businesses in which Westcoast engages.

There is a larger potential in Canada, where provincial Crown corporations own the biggest power utilities. The pot of gold is privatization—the decision by provincial governments to engage the private sector in building all future generating capacity. And the biggest prize of all is in Ontario, where Westcoast has a formidable competitive presence through Union Gas. There are two obstacles to overcome. One is the inertia of the process; privatization is unfolding slowly and is slowed further by public resistance. The second is the large number of players who are competing for a piece of the pie. The outcome depends as much on political skill—honed in the pipeline business, where every decision is three-quarters politics—as on the company's amply demonstrated skill as a power project developer and customer-driven energy provider.

A FINAL PIECE of the puzzle needed to be filled in to complete Westcoast Energy's profile as a fully developed, North American natural gas transportation powerhouse. The company had to have a major, competitive gas marketing operation to provide the services that customers expected in the world of light-handed regulation and to keep the pipeline full.

The company's original marketing operation, involving gas supply contracts with its major Canadian and American utility customers, had been an important but dull business that attracted attention only when it created a crisis. Westcoast had gotten out of the business, not unhappily, at the behest of the British Columbia government and producers in the 1970s.

The acquisitions of Inter-City Gas in 1989 and Union Gas in 1992 included associated gas marketing operations that were the norm for distribution enterprises. In 1993, Mike Phelps recruited executive vice-president Michael Stewart to amalgamate and restructure the two sales operations. Stewart is a native Albertan, born at Okotoks near Calgary and educated as a geologist at Queen's University in Kingston, Ontario. After graduating, he returned to Calgary to work for Canterra Energy, where he became an expert in corporate development and commodity marketing. When Canterra was acquired by Husky Oil, Stewart became an entrepreneur and participated in creating two marketing companies, CanWest Gas Supply and Prism Sulphur Corporation. He became Prism's CEO until Mike Phelps hired him away to Vancouver.

At Westcoast, Michael Stewart merged Unigas and Canadian Hydrocarbons Marketing into Westcoast Gas Services in 1993 and spent the next three years fighting the decline of income margins because of low natural gas prices and stiff competition for customers. The solution, the company decided, was to find a partnership and to create a major gas marketing operation. In 1997, Westcoast partnered with Coastal Corporation to form Engage Energy, a Houston-based North American natural gas and electric power energy trading and marketing company.

"We've had bumps in the road," Stewart said, describing the constant battle to turn wholesale trading and marketing into a more profitable business. In spite of its challenges, Engage Energy has emerged as the strongest Canadian-connected player among its competitors and has fostered a reputation for successful customer relationships. Higher gas and electricity prices have improved its profit potential.

Westcoast, meanwhile, launched two other service start-ups. Enlogix Group was created to sell Westcoast's customer service know-how and technology. It develops, implements and operates computerized billing systems, and quickly established itself as one of the largest North American competitors in the business, with seven clients having a total of 2.2 million customers.

AS THEY SCOPED OUT its future at the end of the twentieth century, Westcoast Energy's strategists had a cornucopia of options. Canadian natural gas was the focus. "The U.S. is the world's biggest natural gas market and continues to grow. Ten years ago, Canada supplied 5 to 7 percent of U.S. gas. Ten years from now it will be 15 percent," director and former chairman Bill Hopper said. "There's sufficient market in the Pacific Northwest to keep Westcoast busy, and there's more gas to be found in the North. The Liard discoveries opened up a new era of supply that will be honey for Westcoast and Alliance. Westcoast owns the low-hanging fruit of the gas business."

The company could remain a pipeliner. It could pursue projects like large pipelines from the Canadian Arctic. The challenge was that it was sometimes a business with relatively low profits—9 to 10 percent return on equity.

Another option was to get into the power generation business as a principal growth focus. But power generation is a lower margin, utility-like business that demands huge amounts of capital and high debt-to-equity ratios, and investors expect it to pay big dividends. It's a treadmill. It doesn't provide exponential growth.

The third option was the service business. But that hadn't proven profitable; quite the opposite, it lost money—not just for Westcoast, but for other pipeline and utility companies who embraced it.

The fourth choice was to move outside the box, as the company did in Mexico, and take its skills to the Third World. The option had its problems. In the $4 billion expansion program involving a dozen construction programs, only Mexico ran behind schedule. It takes a long time to establish business and government relations in a new country, and it would take care to ensure that Westcoast was in the right place at the right time

There was a fifth option: Westcoast could make a big acquisition or initiate a merger that would elevate it to a completely different level.

Those were opportunity-driven moves, and the other side of the coin was that there might be hunters who considered Westcoast as fair prey.

Westcoast Energy crossed the millennium looking more and more like the company Frank McMahon dreamed of in 1934 when he planted the first seeds of its future on the banks of the Pouce Coupe River. It was a continental energy company that had realized, through Alliance Pipeline and Maritimes & Northeast Pipeline, the west-to-east continental span that McMahon envisioned. It had integrated horizontally into electric power. And it was customer-driven, prospering because of its services in a way that profoundly echoed the customer relationships that McMahon assembled to build the transmission line.

There were, however, two larger, intertwined forces in North American energy that were poised to intervene. There was a single, continental energy market—one that Westcoast, as much as any Canadian company, had worked to create. And corporate consolidation—through mergers, acquisitions, partnerships and alliances—was the flavour of the decade.

20

On the Threshold

We are building for the future on an unprecedented scale.
We have reinvested our income. Now is the time to benefit
from the fruits of this investment.
MICHAEL PHELPS

When Westcoast Energy's board of directors sat down for business at their summer meeting on July 27, 2000, they had a remarkable fiefdom to survey. The company was in the home stretch of a three-year, $4 billion capital spending campaign that included the two biggest pipeline projects in North America. It also included one of the most successful foreign investments in Mexican history and a list of power projects that ranked the company among Canadian natural gas–fuelled electricity champions. As the company reached the finish line of its expansion cycle, it wasn't slowing down. The $1.3 billion capital budget for the year was the second largest in company history, and anything but parsimonious even when compared to the 1999 record of $1.6 billion.

The big numbers belied the discipline that maintained, as a matter of policy, the company's A-minus credit rating, in spite of borrowing $7 billion and raising $3.5 billion in equity since 1990. Westcoast Energy had turned in highest-ever annual earnings in 1999; $222 million and

$1.95 a share; the financial results for the first six months of 2000 were also the best in history, at $188 million and $1.63 a share. Clearly, year-end results would handily break the 1999 record, and the company was outperforming its competitors.

The good news had been picked up by financial markets, and Westcoast's common shares were trading at a fifty-two-week high even though pipeline and utility stocks were overshadowed by sexier dot.com-type IPOs and technology plays. By the numbers, Westcoast Energy was at the zenith of its capability, earning the dividend for the hard and dangerous journey it had been on since breaking out over the mountains in 1988 to establish itself as a North American natural gas player.

The directors reviewed a staggering list of projects that were about to start throwing off cash to repay their capital costs and swell the company treasury. The Maritimes & Northeast Pipeline went into service, on schedule, in the fourth quarter of 1999. The Shanghai Power Project was just coming on line, and within months, the Mexican Cantarell Nitrogen and Campeche Compression projects were scheduled to go into operation, as was the Island Cogeneration plant at Elk Falls, on Vancouver Island. Before the end of the year, the interrelated Alliance Pipeline, Vector Pipeline and Aux Sable plant would start delivering gas and gas liquids to customers. In September 2001, the Bayside Power Plant at Saint John, New Brunswick, would go into service. The Indonesian PT Puncakjaya Power had turned in another stellar performance and had been expanded to 388 megawatts of capacity. And southeast of the city of Tacoma, Washington, the company was completing construction of the U.S. $260 million, 249-megawatt Frederickson Power Project.

The Frederickson project wasn't the only indication that Westcoast still had a lot of momentum, even though Michael Phelps was at pains to tell all who would listen that 2000 was a year to digest new projects and make sure that they were meeting their goals. Any asset that doesn't meet expectations is a candidate to be sold, he said at the April annual meeting. The performance bar was high; the company was firing on all sixteen cylinders. In July 1999, Westcoast made an after-tax profit of $59 million by selling the Centra Gas Manitoba gas distribution utility after it became apparent that Manitoba's regulatory authorities seemed unwilling to provide the opportunity to earn the cost of capital. In April 2000, it made $8 million from the sale of its undeveloped Australian natural gas development projects. Over the year the company had also sold off some odds-and-ends businesses that were too small for it to run efficiently.

"A few years ago, we were a British Columbia–based company with almost all of our assets represented by our British Columbia pipeline," Phelps told shareholders at the April 26, 2000, annual meeting. "Today we are a $12 billion North American energy services provider, stretching from Port Hardy, British Columbia, on the West Coast, to Goldboro, Nova Scotia, on the East Coast, and from Fort Liard in the Northwest Territories to the Bay of Campeche in Mexico. We now have a North American footprint."

But Westcoast Energy had not yet exhausted its potential.

"We believe opportunities abound in the natural gas transportation and distribution business," Phelps told shareholders. "Natural gas remains the fuel of choice today and for the future. Governments around the world are grappling with their commitments to sustainable development, and trying to meet emission reduction targets while sustaining economic growth."

Phelps explained that coal and nuclear energy were not expected to be used to meet the needs for new power generation. Oil faced environmental restrictions, price volatility and higher relative costs.

"Natural gas has rapidly become the fuel of choice for broad North American energy needs," he said, explaining that with demand strong and prices at historic highs, Westcoast's role was crucial to meeting the increasing need for natural gas, because it provided the vital link to the two newest natural gas supply regions in North America—northern B.C. and the Northwest Territories at one end of the continent, and the Scotian Shelf at the other.

"Union Gas continues to be an anchor asset in the important distribution franchise area of southwestern Ontario, also owning a well-positioned and very large storage business and controlling the major Canadian transportation hub at Dawn, Ontario," he said.

"We have the people who have been instrumental in the successful development and construction of two of the biggest multi-billion-dollar pipelines in North America," Phelps concluded. "We will not rest here. We will break out of the pack. We still see opportunities out there. We will engage in selective new capital spending when circumstances warrant."

Graham Wilson had his own inimitable way of describing things to shareholders. "Last year was demanding and it's far from over. But we see the light at the end of the tunnel. The bulk of new projects are in the later stages of their construction timetables, and our new energy services businesses are getting their initial problems behind them and looking forward to brighter days ahead. The impact on earnings will be positive on both counts. We are all looking ahead with a strong sense of optimism. Not only are we substantially bigger than a few years ago, the new investment has gone into projects that position us to play an important role in the North American energy market in the future."

BEFORE MICHAEL PHELPS'S fourteen-year stewardship, Westcoast Transmission was a regional company, nonetheless remarkable, but built solely on three pillars: British Columbia natural gas development and transportation, penetration of the province's gas market and exports to the United States. Under Phelps and the team he created, Westcoast became a national story about Canadians making their way successfully in the wider world of the global energy business.

The 1990s were the company's remarkable decade, marked by a rate of investment and growth that would have staggered even Frank McMahon's wildcatter imagination, a decade during which the company raised close to $9 billion in debt and equity capital. That is a number nearly equal to the total annual investment in oil and gas exploration and production activities in Alberta.

Raising the money was a phenomenal achievement, but it was how the money was spent that made the difference. In a stunning array of projects that tested the nerve of the company's shareholders, directors and employees alike, Westcoast pulled off a long list of impossible challenges. It completed the British Columbia gas market by building the Vancouver Island pipeline. It captured two of Canada's oldest and most successful gas distribution utilities and merged them into a natural gas powerhouse. It went to Mexico and promptly became the most successful Canadian company in the post-NAFTA friendly corporate invasion of that country. It extended its global arm to power projects in China and Indonesia.

It extended its continental footprint to Canada's Atlantic coast, where it won the political competition to be the pipeline for Nova Scotia's offshore gas. It joined the Alliance Pipeline consortium as its largest industry partner, putting its shoulder to the wheel to complete that project which, in the words of *The Financial Post,* "has changed Canadian natural gas forever." It expanded and extended its ownership of natural gas–fuelled electric power projects. It marketed its management and marketing expertise to other utilities and created one of the strongest natural gas and electric power marketing operations on the continent.

Michael Phelps's goal at Westcoast was to expand the company's continental footprint—in both depth and reach. When he joined the firm in 1982, it employed 1000 people, had annual revenues of $1.3 billion and earned a profit of $63 million. By the turn of the century, the firm was employing nearly 5100, enjoyed annual revenues of $63 billion, earned $500 million and commanded assets of $15 billion. He intended to double the company's size in the next five years—which may sound improbable, except that this was precisely what Michael Phelps had succeeded in doing over the previous five years . . . and the five years before that.

As Westcoast's chief business strategist, he set the tone and direction of the company and assembled the team to reach its destination. His initiatives turned a sleepy regional pipeline franchise into a great sprawling Canadian multinational, ranked nationally in the top twenty in assets. The company became a continental energy player with its eye on global horizons. The measure of his leadership, however, had not yet been taken because his best projects were just beginning to produce results.

At times unnoticed elsewhere in the country and often unwelcome when they were, the Michael Phelps team, wrote one of the most important energy decades in Canadian business history. These men and their company finally made Vancouver what Frank McMahon always wanted it to be, a ranking city in Canada's energy sector. From the raw salt air, the silent mountains and the impregnable forests, Westcoast Energy had raised the flag of continental greatness.

Canadian natural gas champions have always worried that sooner or later anything they did worth doing would be swallowed up by the ever-expanding, ever-consolidating behemoths of the U.S. lower forty-eight states. In the era of globalization, consolidation, NAFTA and the pre-eminent right of shareholders to gain a quick return on their investment through mergers and takeovers, Westcoast was a prize for any acquisitor.

More than fifty years ago, when Westcoast opened its first bank account, most Canadians heated their homes and cooked their meals over coal and wood fires. Canada was a nation of farmers and farmers' children. Its roads were mostly unpaved, its skies empty of aircraft. Canadians loved the land but had hardly any idea of what it was capable of producing.

And then, with their own hands finally shorn of colonial shackles and brimming with confidence earned by the blood of fathers in war and through Canadian corporate instruments like Westcoast, Canada transformed the northern half of a continent.

It was a process that had its laments. The First Nations were unjustly and unimaginably excluded from the prosperity and equity that were, otherwise, a nation's hallmark. It took a long time to learn how to treat the fragile ecology of the treasured wilderness. Canada gradually became a great nation, and that was as much and more the work of the hands of Westcoast Energy's people as it was of any Ottawa politician or New York financier. This is a company that built

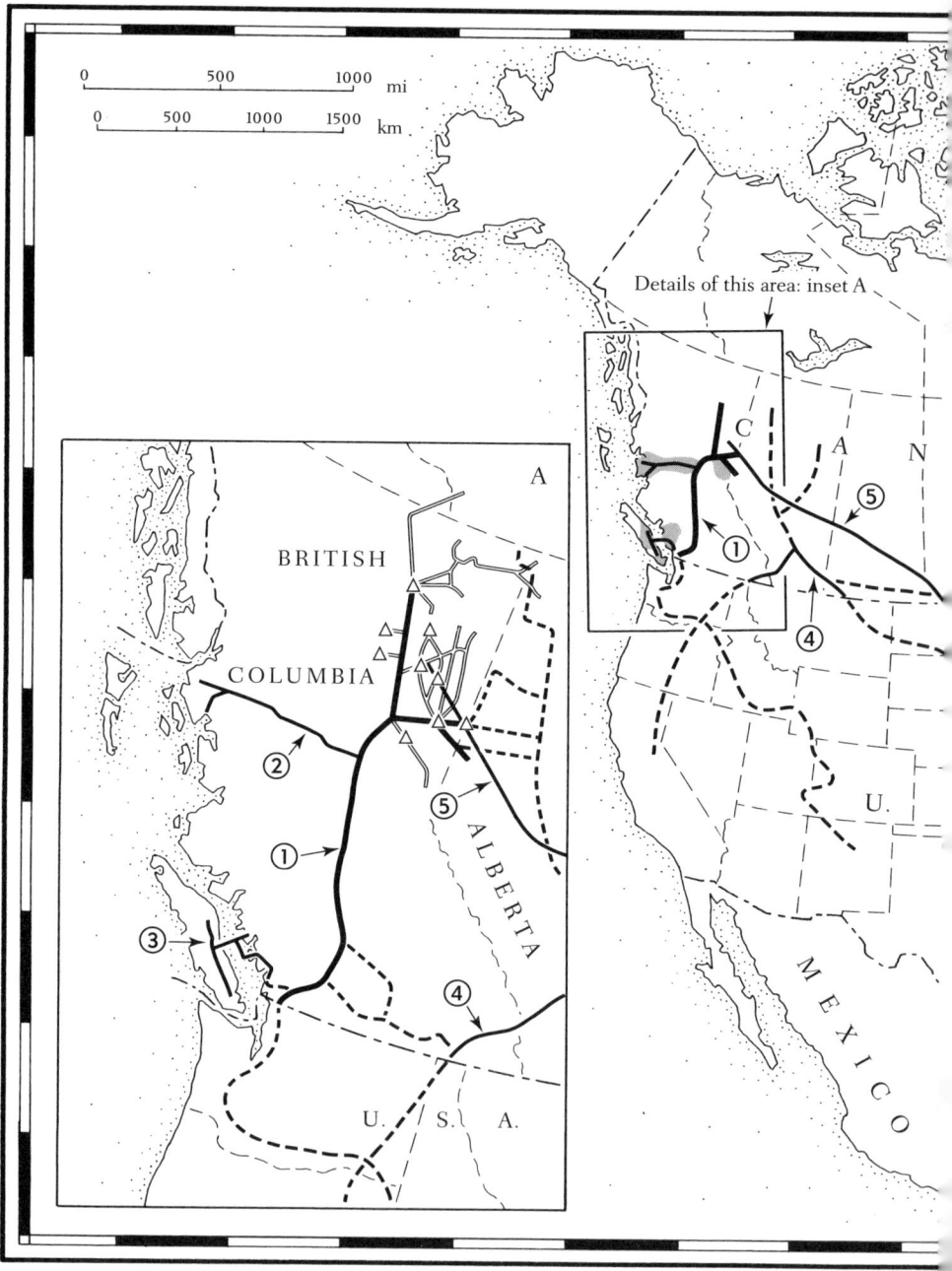

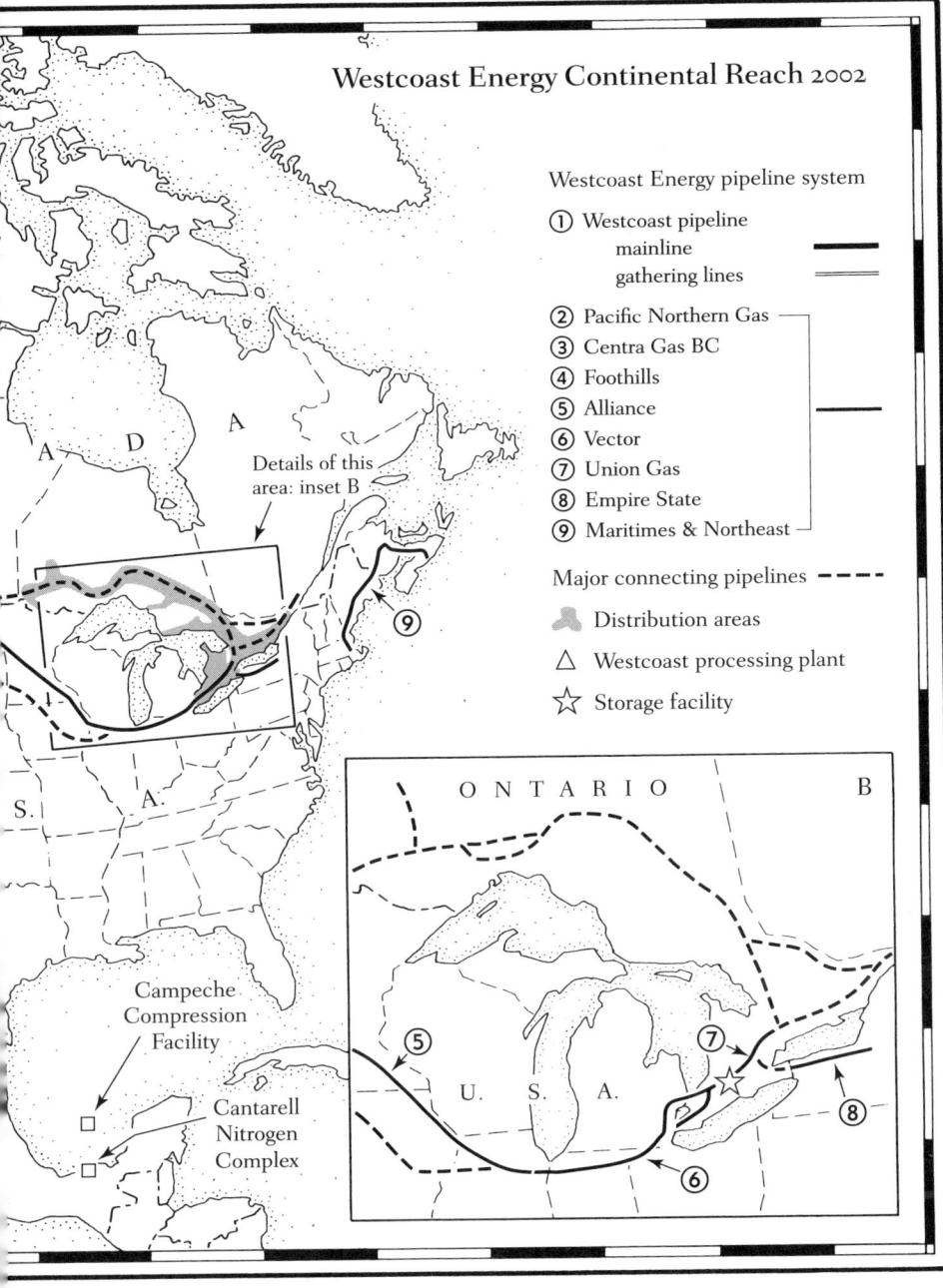

one length of pipeline at a time by the welders of the original Westcoast main line, and they are the icons of its achievements. The poet T. S. Eliot memorably said, "We shall not cease from exploration and the end of all our exploring will be to arrive where we started and know the place for the first time." Thus it is that on the rights-of-way of Westcoast's pipelines and at the power plant fences, it was at last able to look at itself and come to understand what a great thing it had accomplished as one people together.

THE FINANCIAL AND OPERATIONAL results achieved in 1999 were repeated in 2000, when net income was $340 million, and in 2001, when it topped $526 million—a winning streak of three consecutive years of record profitability.

Every new facility in the expansion program proved that it had been worth the risk. Westcoast increased its stake in the Cantarell nitrogen plant by 10 percent and the Campeche Natural Gas Compression plant by 5 percent. Engage Energy became the most profitable merchant gas and merchant power marketing operation of its class. The Bayside Power Project went into service; construction at the Frederickson Power Project remained on track for a summer, 2002 start-up and Westcoast began to plan installation of a second turbine to increase capacity. Meanwhile, the company sold Centra Gas B.C. and Centra Gas Whistler to BC Gas for $590 million. By the numbers, the company was successfully digesting its expansion program and keeping its house in order.

Even as Westcoast Energy's rapid expansion reached its peak, Mike Phelps knew that its success would have an inevitable outcome; the company would find itself to be, as he described it, "a $15 billion company in a $30 billion world." To continue to flourish meant taking on bigger and bigger projects—most notably on the North American energy agenda, the transportation systems needed to get Alaska's North Slopes and Canada's Mackenzie Delta gas out of the Arctic. To play on that field meant accessing very large amounts of capital on competitive terms—two, three, maybe four billion dollars just to get to the table at which the big boys would form up the partnerships to build the systems. A company of Westcoast's size—even with the enlarged balance sheet and increased revenues and income from its expansion—would find it very tough to stay in the game with competitors five or six times its size.

Mike Phelps faced two choices.

He and his company could attempt to reinvent the comfortable caution of the 1960s and 1970s, with the boundary of the new insularity demarcated by its present asset base, instead of the Rocky Mountains of the Westcoast Transmission era. The brutal truth was that the world wouldn't allow such a thing to happen; the company's success pre-empted that possibility and all but guaranteed that a stalker from among the increasingly concentrated and powerful energy transmission, marketing, electric power and services conglomerates would seek out Westcoast and acquire it.

The second choice was to find a partner—in a merger or acquisition—with whose combined strengths Westcoast could stay in the hunt for the big new opportunities of the twenty-first century. In the spirit of Rudyard Kipling's famous description of the predatory character of the wilderness he had seen as "nature, red in tooth and claw," Westcoast could eat or be eaten. Phelps could control the company's destiny by seeking out a deal, or let the company drift until the inevitable acquisitor sought it out and swallowed it. "The circle of CEOs of companies in our business is very small—about ten or fifteen at most," he said. "They are largely American. I started to spend more time with them. I learned to play golf for the first time in my life. There were expressions of interest, but much of it was tire-kicking. When it wasn't serious, I preferred to drop it and say, 'Let's just have a drink.' It is a long, slow dance. We all have companies to run. We all have busy schedules to coordinate. When you start talking, you have to be discreet."

Within the confines of the club where CEOs keep each other's secrets, Phelps's increasingly active social calendar, the number of industry association events and conferences at which he turned up, sent a clear signal to his discerning peers. Westcoast was out to get married.

To all appearances, as 1999 turned into 2000 and was succeeded by 2001, the dance went on, very much behind the scenes. Westcoast was spoken for, even if the arrangement was as tentative as it was confidential.

"It started in the rain, in San Antonio, Texas at an INGA [Interstate Natural Gas Association of America] meeting," Phelps recalled. "We were in a hospitality tent at the edge of a golf course. A couple of Duke executives approached me. They expressed their view that the union of our two companies would create a true continental winner—no one could touch it in terms of supply and market reach. We had worked closely with Duke and got to know them pretty well on the Maritimes & Northeast project, the Alliance Pipeline and our Indonesian power project. While we were of different sizes, we shared similar values and similar cultures. We saw the world the same way. So we started to talk."

They talked for more than a year, finding common ground, narrowing the issues, closing in on a deal.

21

The Deal of a Lifetime

To make an end is to make a beginning,
The end is where we start from.
T. S. ELIOT

The announcement, on September 20, 2001, was crisp and in the context of its impact, brief and to the point. The news release was datelined Charlotte, N.C., and Vancouver, B.C. Its first two paragraphs neatly summed up the story:

"Duke Energy (NYSE:DUK) today announced plans to greatly expand its position in the North American natural gas marketplace by acquiring Westcoast Energy (TSE:W; NYSE:WE), a leading energy company with a significant network of Canadian-based assets, in a cash and stock transaction valued at approximately U.S. $8.5 billion, including debt assumed.

"The transaction provides for the acquisition of all outstanding common shares of Westcoast Energy in exchange for a combination of cash, Duke Energy common shares and exchangeable shares of a Canadian subsidiary of Duke Energy such that 50 percent of the consideration will be paid in cash and 50 percent will be paid in stock. The transaction is intended to provide Westcoast shareholders with approximately Cdn$43.80 per share in value. Duke Energy believes that the acquisition can be completed during first quarter 2002."

On WEnet, Westcoast's intranet system, employees found a posting from CEO Michael Phelps that opened: "Westcoast Energy's Board of Directors wishes to announce that it has approved the acceptance of a proposal by Duke Energy Ltd. for a combination with Westcoast Energy Inc. and has recommended the acceptance of the proposal by the shareholders."

A front-page story in *The Financial Post* on the following morning, headlined "An Ideal Marriage," quoted Michael Phelps: "The continental energy marketplace has been a fact for some years, but it is beginning to have a very good ride. We'll be worried about the short-term economy, but we have a high level of confidence about the growth requirement for energy infrastructure and especially for the role of natural gas in that economy. If you believe in that story, the ideal marriage is one that has an existing solid U.S. infrastructure platform, sits beside premium markets and marries Canadian infrastructure and access to Canadian supplies."

There had been the usual stirrings and speculation. The Canadian oil and natural gas sector was undergoing a year of consolidation that would end with nearly $35 billion in takeovers just counting those involving American companies acquiring their Canadian counterparts and nearly ten times that ($250 billion) in energy mergers and acquisitions around the world. A $13.6 billion acquisition involves a large cast of employees and advisors whose activities, especially in the final period of negotiations, are difficult to mask. The Westcoast-Duke deal had taken nearly two years to complete, with the chances of a leak rising with each passing month. In the autumn of 2001, there were just three large-scale Canadian pipeline companies—Westcoast, TransCanada PipeLines and Enbridge Energy—and how they would fare, as their North American sector fused into a smaller number of very large players, was a staple of street talk.

On September 11, terrorists destroyed New York's World Trade Center, a portion of the Pentagon and a commercial jet in flight over Pennsylvania, killing more than 3000 people, most of whom worked at the heart of global financial and energy markets, and shutting down the

New York Stock Exchange. In late September the North American business community was preoccupied and overwhelmed by the impact and consequences of the horrific event. The chill that the attacks on New York and Washington brought deepened the spreading gloom about a recession in Canada and the U.S. Sliding natural gas and oil prices cut into energy-sector profits and put the brakes on new pipeline and electric power generation projects.

In the universal scheme of things, the Duke and Westcoast announcement may not have been earthshaking and for many, perhaps most, of Westcoast's shareholders it might not have been surprising. It was, however, life-changing for the company's employees. Some would become wealthy through it; many would stay and trade in Westcoast's past for Duke's future. Some would step out into the wider world and find new successes, some would find it a difficult struggle to move on, and some would quietly end their careers in retirement. None would forget having been a part of the high-octane years during which Westcoast went from the cautious regional utility to the aggressive Canadian adventurer with a continental reach.

The deal had landmark implications for the brutally competitive energy pipeline, marketing and trading, power and services business in North America. When the transaction closed, six months later in the spring of 2002, after all of the paperwork and approvals were complete and the last of the cheques cut, Duke Energy Ltd. emerged as the largest and strongest energy transmission, power generation and energy services company on the continent and one of the largest in the world.

Before the combination, Duke's footprint of pipelines, power plants, storage and processing facilities, and service offices was concentrated in the eastern and southern United States, south of a line drawn from Nova Scotia to Utah. With Westcoast, Duke filled in its North American map and became the only member of its class with pipeline and power assets across the continent, including an inside track on future pipelines and gas storage needed to bring Alaska's North Slope gas to markets in the lower forty-eight states and a foothold with owned assets in Mexico.

Duke and Westcoast combined created a new class of energy company—a $68 billion powerhouse with annual earnings before tax, using 2000 figures, of some $3 billion, and a uniquely balanced portfolio of assets, with gas pipelines and storage making up 22 percent of the assets; franchised electricity generation, 39 percent; energy trading, marketing and consulting services, 30 percent; and field services, 9 percent.

A CONTINENT APART, with about 4000 kilometres separating their respective headquarters, Westcoast Energy and Duke Energy belong to two separate worlds. Both come from communities that are the products of regional geography and history, and there are some parallels between these elements. The differences, however, prevail.

While both are coastal regions backed by mountains, British Columbia and Vancouver are, by comparison to the Carolinas and Charlotte, young, raw and untested. European contact worth mentioning commenced in British Columbia in the 1770s when Spanish explorer José Maria Narvaez and British Captain George Vancouver probed the coast, and mapmaker Alexander Mackenzie traversed the Peace River country and reached the Pacific Coast from the continent in 1793. Vancouver's first permanent settlement dates back to just 1820. In contrast, Sir Walter Raleigh made the first modern European attempts to settle North America at Roanoke Island, part of present-day North Carolina's Outer Banks, in the late 1580s. The city of Charlotte was settled in 1750 and incorporated in 1768.

Although Duke Energy's head office is in North Carolina, its history belongs as much to South Carolina, and the states share a common geography, with a long, irregular Atlantic coast and a lush coastal plain that makes up two-thirds of South and nearly half of North Carolina's land area. The plain rises to the Piedmont plateau, a rolling, hilly upland laced with rivers and dotted with great swamps. Behind the plateau tower the Appalachians, bearing the historic names of the Blue Ridge and Great Smokey and Sassafras Mountains.

Just as Westcoast Energy was founded on the geology of interior British Columbia, in the form of the rich gas fields of the Peace River country, so Duke owes its origin to the rivers that pour out of the mountains and rush southeastward down to the Atlantic Ocean, parachuting over the Piedmont fall line, a sharp drop in elevation that divides the plateau and the plains. Those rivers are among the lower forty-eight states' best natural hydroelectric power producers and have become among the most intensively developed for that purpose on the continent. Exploiting hydropower potential was Duke's first business.

There is one important similarity between British Columbia and the Carolinas. Both are fiercely independent and have a history of dissent and conflict with their respective federal governments. One of the first protests against the British during the American Revolution, the Mecklenburg Declaration of Independence, was signed at Charlotte in 1775, a year before the U.S. Declaration of Independence. Thereafter the British referred to the city as "that nest of rebellion."

The Carolinas were included in the founding thirteen states to sign the Declaration of Independence drafted by Thomas Jefferson and adopted by the Continental Congress. It is a matter of continuing pride to their citizens that South Carolina was the eighth state and North Carolina the twelfth to sign the 1776 Declaration, but they did so on the strength of support from seaboard communities and in spite of dissent from the back-country Piedmont and Appalachian voters. Subsequently, the two were among the most reluctant new states to cede rights given by the British to the former colonies to western lands that ran to the Mississippi River. It wasn't until 1787, when the U.S. Constitution was adopted, that South Carolina gave up its western land rights, and it was 1790 when North Carolina followed.

South Carolina was the first Confederate state to secede from the Union, on December 1860. North Carolina was one of the most reluctant, seeking a resolution to the impending conflict until May 1861; but when North Carolina joined the cause, it did so without reservation and its soldiers accounted for more than forty thousand casualties— one-quarter of the losses of the Confederacy.

Although British Columbia's reluctance to join Canadian Confederation, holding back four years until a transcontinental railway was promised, contrasts with the Carolina's founding support for the American Union, Carolinians of the eighteenth century would have empathized with B.C.'s threat to reconsider and secede when completion of the railway was delayed until 1876. Equally, Carolina's founders would readily understand the sentiments that have nurtured Western Canadian protest movements—including the CCF and NDP, Social Credit and the Reform Party.

The Civil War and the so-called reconstruction devastated the economies of North and South Carolina. Like the rest of South, they were excluded from the 1890s Gilded Age, the first industrial boom in American history—except to the extent they were plundered by the Northern carpetbaggers. Rich in tobacco, indigo, timber, quarry stone, rice and cotton, they could not be forever denied. After the turn of the century, things began to look up with the development of textile and chemical industries. The Carolina industrial advantage was hydroelectric power.

Duke Energy traces its corporate beginnings to James Buchanan Duke, who founded the American Development Company and acquired land and water rights in 1899 on the Catawba River on the border country joining the two states. In a striking parallel to the McMahons, the elder Duke's partner was his brother, Benjamin Newton Duke. Within six years, they had joined up to form the Southern Power Company with another pair of brothers, Dr. Walker Gill Wiley and Dr. Robert H. Wiley of the Catawba Power Company, which in 1900 had started construction of a hydro station on the Catawba, at India Hook shoals. The plant was to generate electricity for Rock Hill, a South Carolina chemical and textile centre on the Catawba Carolina, five miles south of the North Carolina border.

The day when the Catawba Hydro Station went into operation—April 30, 1904—stands as the birthdate of the Duke Power Company, and of Duke Energy. The Catawba must be the most unassuming river on the North American electric power grid. It is really just a North Carolinian headwater for a South Carolina river that changes its name twice, to the Wateree and then to the Santee, before it flows into the

Atlantic Ocean. Nevertheless, by 1930, shortly after Duke Power Company and Southern Power merged, the company had built ten more hydropower stations on the river, at Oxford, Cedar Creek, Rhodhiss, Mountain Island, Dearborn, Wateree, Fishing Creek, Lookout Shoals, Rocky Creek and Great Falls. Water from the river was employed in other coal-fired steam plants.

Over the next sixty-six years, until the 1996 merger with PanEnergy Corporation of Houston transformed Duke from an electric utility to an energy company, appropriately renamed Duke Energy Ltd., with major natural gas pipeline, processing and field services assets and a budding energy services business, Duke Power grew steadily and at times aggressively, joining the nuclear power club in the mid-1950s while continuing to build new coal-fired steam and hydroelectric plants and transmission infrastructure at a steady rate.

When Duke made its formal, public offer to acquire Westcoast Energy on September 20, 2001, it ranked in the top ten of world energy companies. It owned worldwide assets worth U.S. $58 billion, including 30,000 megawatts of electric power generation and 13,000 miles of operated natural gas pipeline. It had 23,000 employees worldwide, with operations and services in fifty countries. It transported 8 percent of the gas burned in the United States. It was the country's largest natural gas liquids producer, second-largest gas marketer, third-largest power generator and a top-ten U.S. electricity producer.

Duke generated revenues in 2000 of U.S. $49 billion, making U.S. $1.8 billion after taxes. It ranked number seventeen on *Fortune*'s list of top U.S. companies. *The Financial Times* of London called it the world's most respected utility, and it ranked number two in the Dow Jones 2000 ranking of utilities by return on equity. Its list of awards and honours for community service, corporate achievement and top-drawer position in a dozen important rankings by performance was as long as some of its power lines.

Westcoast was, by comparison, one-sixth the size and not as well known, outside Canada or Mexico. The magic of the deal was, to steal a phrase from Shakespeare, a marriage of true minds; what the two companies had in common had little to do with size and everything to do with character, style and vision.

THE 1996 PANENERGY–DUKE POWER merger brought into the new Duke Energy a collection of well-developed, mature and profitable natural gas pipeline, processing and field service assets that ranked PanEnergy with the American majors. In the months that followed, Duke's chairman, president and CEO, Richard Priory, who had held his job for just two years, became acquainted with Westcoast chairman and CEO Michael Phelps. The two became business partners in a power plant in Indonesia, the Alliance Pipeline project and, perhaps most important in cementing the relationship, the Maritimes & Northeast Pipeline project. From 1999 on, they talked regularly and with increasing seriousness about a combination of their two companies.

For Duke, the reason was straightforward. The company was on a fast track. It had become a national and international energy powerhouse. As Duke told its analysts, there wasn't a better package of well-thought-out, profitable and promising assets to be had in North America, or anywhere in the world for that matter, which fit as well with the company's growth initiatives. The deal would advance the company's corporate strategies and financial goals; it would be a major step in Duke's growth and evolution. "This is a company we respect and know well," Richard Priory told the market. "We are combining to realize the growth potential that both have created."

As CEO of the smaller of the two companies, and the one that would be absorbed in the transaction, Michael Phelps had another side of the story to tell his employees on September 20: "Given the size and financial strength of our competitors—how do we capture these opportunities and access sufficient capital to continue our growth? Through the deal, our shareholders will realize the value that has been created as well as having the opportunity to participate in the prospects of the combined companies. As we go forward, the transaction gives us the size, the financial strength and the depth to pursue opportunities that we would be challenged to efficiently pursue on our own. Our two companies are a good fit; in fact, I believe we are the best fit for each other," Phelps said. As the weeks passed during which the companies dotted the i's, cross the t's and gained shareholder, creditor and regulatory approvals, Phelps repeatedly made the point to many of his constituencies that this deal rewarded the shareholders generously and appropriately for the risks they had taken in the investments they had made in the company for so many years.

THERE WAS ONE FINAL golden moment for Michael Phelps, his board and the team of men and women who had taken Westcoast Energy to its greatest heights, then won a deal that rewarded them and their shareholders, guaranteeing the future of their projects and dreams. On February 13, 2002, the company announced that its last year of operation, ending December 31, 2001, had produced a record net profit of $526 million, or $4.26 per common share. Not only was it the company's best-ever annual result, it topped the previous year by 55 percent, making it one of the best-performing public companies not just in the energy sector but in Canada. It was a glimmer of good news in the market, in a year in which the economy was struggling to stay above water.

"It was a ninth-inning home run," Michael Phelps said. It demonstrated the wisdom of Duke's decision to combine the companies, and of Westcoast's shareholders to endorse the deal nearly unanimously.

"We have built a company that has rewarded its shareholders with a steady stream of dividends and increased share valuation. This is the direct result of a strategy that has been implemented by a superb team of employees," Phelps said when announcing the profit. "It is fitting that we conclude Westcoast's public trading on both an outstanding record of financial performance and the overwhelming 96 percent vote of our shareholders to endorse the board of directors' recommendation in favour of the acquisition by Duke Energy."

After the final financial results in the history of Westcoast Energy were released and the board had met for the last time to vote a final quarterly dividend of 34 cents per common share, an inner circle of directors and past and present executives met for one final dinner together. It was a moment of civility and grace as well as celebration. There had been many such dinners over the years to mark the milestones since the glittering ball forty-five years before in the Hotel Vancouver to mark the construction and launch of service on the Westcoast Transmission pipeline. But this was the last chapter.

Three generations of leadership had come and gone in the interval, and no one at the final dinner had been at the first. But the air was as heavy with memory as it was charged with possibility. Frank McMahon had no idea that his creation, too often held together with the baling wire and binder twine of corporate expediency, would come to this.

The chief raconteurs in the room, Ed Phillips and Bill Hopper, rose to tell stories that evoked the years flown past. Michael Phelps—ever the measured, private man—spoke to the numbers that marked the shared success of all in the room, how much had been built and how well the shareholders had been rewarded.

Then, reluctantly and with deeper feelings than they, being members of a national corporate culture that treasured its reserve more than its bank balances, would express openly, they went out into the night.

A company dies, formally, when the closing papers of an acquisition are signed—and that event was still several weeks ahead. In human terms, however, Westcoast Energy more than most Canadian companies had been a human experience as well as a successful business; in human terms what had been created in the heat of the wild well flare on the Pouce Coupe River came to its dignified end in the cool, damp Vancouver night beneath gleaming shards of urban winter light.

Index

Alberta Distillers Ltd., 36–37
Alberta Energy Company, 171, 173, 199
Alberta Gas Trunk Line (AGTL), 81, 96–99, 100, 113
Alcan/Foothills. *see* Foothills Pipe Lines
Alliance Pipeline: beginnings, 198, 200–201; financial backers, 199; helping Fort St. John, 189; Union Gas merger, 165
Anderson, John, 181; biography, 82, 114–15; expansion, recessions, growth, 129–30, 135; illness, death and legacy, 130–32, 135; strategic plan, 132; takeover, the aftermath, 115–16, 117; taking over at Westcoast, 125–27
Arctic Natural Gas Transportation System, 213
Australia pipeline project, 225–26
Aux Sable Liquid Products, 200, 201

Barrett, Premier Dave, 101, 102, 103–4
B.C. Endowment Fund, 149
B.C. Petroleum Corporation (BCPC), 115, 122
BC Gas, 174
BC Hydro, 103, 122–23, 170
Bear Oil Co., 28
Bechtel, Steve, 50, 168
Behn, Chief George, 218

Bell, Max, 36
Bennett, Premier Bill, 114, 121–24, 149, 170
Berger Royal Commission, 99–100
Blair, Bob, 82, 97, 98, 99, 113
BOC Gases, 205, 206
Bodnar, Bohdan, 143, 184
Borden Royal Commission on Energy, 71–73
Bouchard, Premier Lucien, 195
Boucher, Doug, 19
Bowsher, Pat, 23–24, 50, 77–78
Brandl, Victor, pipeliner, 46
British Columbia government. *see* Government of British Columbia
Brown, Bobby, Jr., 22, 36

Campeche pipeline project, 210
Canadian Association of Petroleum Producers, 195, 219
Canadian Bechtel Limited, 35
Canadian Development Corporation (CDC), 109
Canadian Football League, 40
Canadian Imperial Bank of Commerce (CIBC), 148
Canadian International Development Agency (CIDA), 207, 208
Canol Oil pipeline, 63–64

Cantarell Nitrogen Project: 206–7, 209, 210; bidding on, 205–6; BOC Gases consortium, 206; environmental review, 207, 210; financing, 208–9; political unrest, 207–8, 209
Chieftain Developments, 169, 170, 171
Chrétien, Prime Minister Jean, 195
Citibank of New York, 208–9
Clark, Karl, 25
Clark, Premier Glen, 171, 189
Columbia Oils Ltd., 19–20, 21–22
Companies created by Frank McMahon. see Bear Oil Co.; Columbia Oils Ltd.; Crowsnest Pass Oils; Gas Trunk Line of British Columbia; Mountain Pacific Pipeline; Pacific Northern Gas; Pacific Petroleums; Saratoga Processing Co.; West Turner Valley Petroleums; Westcoast Petroleum; Westcoast Transmission Company; Western Products and Crude Oil Pipeline Company
Computerized billing system, 229
Coned, 103
Corporate culture, post-1996, a traumatic shift, 180, 182–85, 215
Corporate culture, pre-1996, cradle-to-grave, 61–62, 181–82
Coste, Denis, 156
Coste, Eugene, 156
Crawford, Jack, 199
Crowsnest Pass Oils, 19; pipeline, 76

Davidson, Ken, 160, 161
Dawn, Ontario, 156–57, 163, 164, 165, 201
Deregulation, 129, 164, 165, 173
Devonian oil field, 27–28
Dickie, Chief Harry, pipeliner, 42, 43, 218–19
Dinning Royal Commission, 31
Duke, James Buchanan, 248
Duke Energy: acquisition of Westcoast Energy, 243–45; company footprint, 245–66; fast tracking, 250; founding and development, 248–49; impact of geography and history, 246–48; Westcoast partner, 192, 197

Ebel, Greg, 143
Edgeworth, Al, 142
El Paso Energy International, 204
El Paso Natural Gas, gas exports, 34, 35, 102, 115
Electric power generation, 228
Engage Energy, 227, 229
Enlogix Group, 229
Environmental concerns: duty of compliance, 215; Foothills Pipe Lines, 221; review process, 207, 210; Vancouver Island pipeline, 176
Ethics in Action Award, 220
Export Development Corporation (Canada), 209

Farris, Ralph, 152–54
First Nations, 194, 211, 221; building relationships, 214, 215, 216
Foothills Pipe Lines, 81, 96–101; discovery and planning, 220–21; economic boom, 222; gathering lines and transportation, 222; LNG technology, 223–24; a partnership, 213; resistance to pipeline, 221
Fort Nelson: building community infrastructure, 188, 218; history, 216–18; relationship with aboriginal people, 216; voluntary taxation, 188, 218
Fung, Bob, 163

Gas Arctic Study. see Foothills Pipe Lines
Gas Trunk Line of British Columbia, 75
Gaz Métropolitain, 193–94
Gesner, Dr, Abraham, 155
Gibson, Kelly, the Enforcer, 79–83, 94, 95, 116; and Petro-Canada, 111–13
Government of Alberta, 141
Government of British Columbia: B.C. not for sale, 121–24; developing Peace Country, 30; oligopoly problems, 101–4; right-of-way, 76–77; and Westcoast Energy, 149; withholding mineral rights, 21–22

INDEX

Government of Canada, and Petro-Canada, 110, 119–21
Graham, Robert G. ("Bob"), 152, 154, 161
Great Canadian Oil Sands Plant, 25
Grizzly Valley processing facility, 174
Guichon, Bernie, pipeliner, 41–42, 44–46
Guilliame, Norman, 19

Haberl, Steve, 199
Hall, Lindsay, 143
Harrold, Bill, 64–66
Harvie, Eric L., 36
Herron, Bill, 36
Hetherington, Dr. Charles, 77; company lore, 35–36; the company's sail, 32–33; pipeline plan, Peace River Country, 77, 123, 169; post Westcoast, 80–81; the Western Loop, 169
Hibernia offshore oil, 148
Home Oil, 22
Hopper, Bill, biography, 118–20; choosing Anderson's successor, 131, 135–36, 139; and Petro-Canada, 111–12, 117, 120–21; and Westcoast, 112–13, 116–18, 125–26, 127, 149
Howe, C.D., 89, 153
Hume, George, 19, 34
Hydro-Québec, 195
Hydrostatic test, 53–54

Imperial Oil: Leduc oil field, 26; Redwater oil field, 27; Westcoast price contracts, 102–3; wildcat well, Pouce Coupe, 4, 21
Inter-City Gas Corp. (ICG), 147; acquired by Westcoast, 159–63; history, 151–55
Irian Jaya Power Plant, 225
Island and Coastal Pipeline, 168

Johnson, Ed, 82

Kavanagh, Jack, 171–72
Koop, Irv, 142, 143, 177, 182–83
Kutney, Peter, 33

Lagadin, John, 199
Lang, Otto, 138

Langan, Pat, 195
Laundry, Al, 186–88
Leduc oil field, 26
Lee, Charles, 36
Leech, Jim, 158
Liberal government, a national pipeline, 195; pipeline debate, 153–54
Light-handed regulation, 176–77
Link, Ted, 28
Liquified natural gas (LNG) terminals, 220–21, 224
Littledale, Dick, 82
Lougheed, Premier Peter, 141
Lund, Dewey, pipeliner, 48, 53

Mackenzie Valley pipeline, 222–23
Magnum Project, 168
Mann, George, 158
Manning, Premier Ernest, 31
Maps: Westcoast Energy Continental Reach, 238–39; Westcoast Transmission System, 60
Maritimes & Northeast Pipeline: application to NEB, 192, 193–95, 197; Maritimes renaissance, 190–91; politics, 192–94, 195; regulatory hearings, 196–97; and Sable Offshore Energy Project pipeline, 192; tolling dispute, 194–97; Trans Quebec and Maritime Pipeline proposal, 194–95
Marriot, Peter, 161
Martin, George, Vancouver investor, 19, 24, 30
McDonald, Douglas Peter ("D.P."): the company's anchor, 33, 77–78; joins Westcoast, 31–32
McIntosh, Al, 113
McKenna, Premier Frank, 197
McKeough, Darcy, 158
McLeod, Joanne, 142
McMahon, Francis Joseph (father): boomtown drifter, 7–9, 12, 15–16
McMahon, Francis Murray Patrick ("Frank"), and arctic gas pipelines, 213; celebration, 58–59, 61; death, 131; Depression wildcatter, 17–20, 22–25;

the dream, 3–5; the early years, 8–13; first jobs, first companies, 13–15; the last battle, 76–78; the later years, 93, 94; the Leduc field, 26–28; lifestyle of a millionaire, 36–39, 40; marriage and family, 15, 16, 37–38; pursuing the dream, 20–22, 30–36; Vancouver Island pipeline and, 168–69
McMahon, Frank, Jr. (son), 16, 37
McMahon, George, (brother), 8–9, 14, 76; the inner circle, 24, 77–78; lifestyle, 39, 40
McMahon, John (brother), 8, 14, 19, 36, 37
McMahon, Marion (daughter), 37
McMahon, Mrs. Francis (mother) (Stella Soper), 8–12, 14
McMahon, Mrs. Frank (wife) (Elizabeth "Betty" Bettes), 37–39, 40
McMahon, Mrs. Frank (wife) (Isobel Grant), 15–16, 37
McMahon, Patrick (uncle), 7–9
McManus, Rob, 189
Merrill, Bob, 82
Millennium Project, 202
Mother Westcoast, 175, 181, 182
Mountain Pacific Pipeline, 91
Moyie, B.C., 13–14, 19
Mulroney, Brian, 129
Murphy, Ron, 69–70
Muskwa-Kechika Management Area: Ethics in Action Award, 220

National Energy Board (NEB), 72, 123, 128, 141, 192, 197, 201; and northeast expansion, 174–78, 193
National Energy Policy, 71–72
National Energy Program, 100, 123, 127
Nationalism, Canadian, 110–11, 195
Natural gas: the Arctic gas fields, 220; exports to the U.S., 34–35, 72; growth in demand, 135; a nuisance by-product, 4; the Pointed Mountain field, 200; the Scotian Shelf, 190, 191
Neville, Bill, 146
Norman Wells oil discovery, 21

North American Free Trade Agreement (NAFTA), 204
North Slope Alaska project, LNG technology, 223–24
Northern and Central Gas, 152, 154, 168
Northern Area Transportation Study, 199
Northern Ontario Natural Gas (NONG), 52, 152, 153–54

Oil and gas policy, federal. *see* Policy, federal oil and gas
Oil industry: effect of McMahon brothers on, 29–30; effect of Second World War on, 24–25, 29; growth by acquisition, 109–10; privatization, 228; the start-up years, 92
Oil sands, 25, 148, 191
OPEC (Organization of Petroleum Exporting Countries), 97, 110
Owen, Doug, 77, 79

Pacific Gas and Electric (PG&E), 102
Pacific Northern Gas, 67; pipeline, 68–70, 92–93
Pacific Northwest Pipeline Corp., 34, 35
Pacific Petroleums: acquisition, 109–13; and aftershock, 113–17; blow-out and success, 26–28; drilling, Peace River, 30, 35; looking for a partner, 73–76; oil sands, 25–26; start-up, 24; and Westcoast, 73, 117–18
Parker, Frank, 181–82
Parkinson, Derek, 127, 128; interim president, 131
Pattulo, Premier Duff, 21–22
Peace River country, 46–48, 92, 214–15
Permack, Harvey, 171, 205
Perry, Glen, 199
Petro-Canada: acquisition and expansion, 110, 112–17, 117–18, 120–21, 147–48; creation, 120–21; Hibernia oil field, 148; Syncrude oil sands, 148; and Westcoast, 127, 148–49
Petroleos Mexicanos (PEMEX), 203–5, 206–7, 209, 210, 211
Petroleum industry: the start-up years, 92

Pew, John Howard, 25, 26, 31
Phelps, Michael ("Mike"): Alliance Pipeline project, 201; Anderson's team, 127, 128; building a team, 141, 142–43; creating a corporate footprint, 146, 234, 236; education and early career years, 136–38; future: merger or acquisition, 240–42; impact on Westcoast, 143–44; leadership style, 144–45, 146; Liberal connections, 137–38; mandate for company growth, 144, 145; Maritimes & Northeast Pipeline project, 196; Mexico, 210–11; Muskwa-Kechika land-use protection area, 219–20; quintessentially Canadian, 136; recruited by Westcoast, 136, 139–40; Westcoast CEO, 133; and Westcoast Petroleum, 146–47
Phillips, Edwin ("Ed"), 149; acquisition, 109–13; and Arctic Natural Gas Transportation System, 213; biography, 87–91; and First Nations, 214; and Foothills Pipe Lines, 99, 100, 213; the money manager, 84–87, 91–95; pipeline crisis, 65, 66; as president, 101, 104–6, 108, 118
Phillips Petroleum, 80; and Westcoast, 74–75, 76
Pickell, Cec, 82
Pigs and pigging, 53
Pipeline projects. *see* Alliance Pipeline; Arctic Natural Gas Transportation System; Campeche; Crownest Pass Oil; Foothills Pipe Lines; Maritimes & Northeast Pipeline; Pacific Northern Gas; Sable Offshore Energy Project; Vancouver Island Natural Gas; Vector Pipeline; Westcoast Transmission; Western Pacific Products and Crude Oil Pipeline; Yoyo
Pointed Mountain gas field, 103
Policy, federal oil and gas, 110–11, 119
Power generation plants, 226
Price, Randy, 143, 161
Prudhoe Bay oil field, 97

Rassmussen, Merrill, 113
Redwater oil field, 27–28
Reid, Bob, 142, 164, 225
Retrutiak, Ken, 142, 143
Richardson, Ernie, 114
Rights, oil and gas, 111
Ross, J. C., 19
Ross, Lieutenant-governor Frank, 19, 36, 61
Royal Bank of Canada, 75
Rutherford, Ron, 70, 98; enterprising engineer, 106–7; Vancouver Island pipeline and, 168, 169, 170

Sable Offshore Energy Project (SOEP), 192, 194, 196–97
Saratoga Processing Co., 76
Savage, Premier John, 197
Schwitzer, Eric, 142–43
Second World War: effect on oil industry, 24–25, 29
Sehmer, Dr. John, 130–31, 145, 184, 185
Shannon, Red, 82
Shippley, C. S., 18–19
Shore, Moody, 23
Smith, Jack, 106, 107–8
Solex Energy gas liquids plant: explosion, 186–89
Soper, Wayne, 143, 216
Sour gas, 54–55
Spencer, Victor, 18–19, 30
Starr, Chuck, 82
Stewart, Michael, 142, 200, 205, 229
Stewart, R. B. ("Bob"), 78–79
Strong, Maurice, 111, 120

Thompson, Owen C., 9, 10, 14, 19
Trans Quebec and Maritime Pipeline (TQM): SOEP proposal, 193–95
Trans-Alaska oil pipeline, 97
TransCanada PipeLines, 33, 71–72, 141, 193–94, 195, 200; the great pipeline debate, 153
Tripp, Charles, 155
Trudeau, Prime Minister Pierre, 110–11
Turner Valley, 15, 22

Unicorp Canada Corporation, 158–59
Union National Gas Company (Union Gas): acquired by Westcoast, 163–66; history, 151, 155–59
Union Natural Gas Company, 156–57
Union Oil, 169
Union Shield Resources, 158
United States Federal Power Commission (FPC), 34, 35
Unruh, David ("Dave"), 142, 143, 159, 183, 193

Vancouver Island Natural Gas pipeline, 122–24, 147, 171; achievement, 172–73; history, 168–72
Vector Pipeline, 165, 201–202

We Change to Win, 174, 180, 183, 185
Wei Gang Power Plant, 227
Wespec pipeline, 78
West Turner Valley Petroleums, 23
Westcoast Energy acquisitions, 146–47; Inter-City Gas, 147, 159–63; Union Gas, 163–66
Westcoast Energy Company: Alberta Energy Company, 147; Alliance Pipeline project, 200–1; Cantarell Nitrogen Project consortium, 206; continental footprint, map, 239; defining moments, 182–83; and Duke Energy, 243–45; final moments, 251–52; financial profile, 232–33, 234–35, 240; and Foothills Pipe Lines, 221; footprint, 204, 234, 236–37; gas pipeline, a first, 222; gas measurement system, 237–38; gas plant explosion, 186–89; global investments, 225–26; growth, 230–31; merger or acquisition, 240–42; Mexico, 203–5, 206–7, 209, 210, 211; offshore energy, 192; and Petro-Canada, 147, 148–49. *see also* Westcoast Energy acquisitions
Westcoast Energy International, 225
Westcoast Petroleum, 46, 116–17, 118, 129; fifteen-year strategic plan, 128–29, 132; growth, 146–47; sale of, 161–62

Westcoast Transmission Company: acquisition and aftershock, 109–13, 112–17; and Alberta Energy Company, 171; a brief history, 1–2; at a crossroads, 134–36; the fifteen-year strategic plan, 128–29; finances, crisis and stability, 64–65, 74–75, 78, 79–83, 93–95; the first pipeline, 33–34; gas export struggles, 34–35; head office, 93; incorporation and the inner circle, 31–33; leadership crisis, 71, 72–74; map of pipeline system, 60; a new direction, 127; oligopoly, 101–2; and Pacific Coast Energy, 171; and Pacific Petroleums, 73, 117–18; the people, 56; pipeliners and producers, 173–78; the Pointed Mountain affair, 103–4; recession, 139; and Vancouver Island pipeline, 122–24; as Westcoast Energy, 133
Westcoast Transmission pipeline: the achievement, 57–58, 63–64; the changes it wrought, 58; construction, 49–51, 52–56; initial steps, 33, 34–36; the pipeliners, 43, 44–45, 49–50 (*see also* Brandl, Victor; Dickie, Harry; Guichon, Bernie)
Western Pacific Products and Crude Oil Pipeline Company (Wespac), 77–78
Whittall, Norm, 19, 30, 36
Wilkinson, Bob, 19
Williams, James, 155
Willms, Arthur ("Art"), 183; Anderson's team, 127–28; baptism of fire, 140–41; executive vice-president, 140; and National Energy Board, 128, 141; Phelp's team, 140, 141, 143; and SOEP rates, 196
Wilson, Graham, financial wizard, 141–42, 148; financing Cantarell project, 208–9; Holy Trinity, 141; Inter-City Gas acquisition, 159–61; report to shareholders, 236
Wyman, Bob, 135

Youell, Len, 33
Yoyo pipeline, 64–67